The Niagara Escarpment

15
3/1995 f

Eden

April 1975

The Niagara Escarpment
From Tobermory to Niagara Falls

William Gillard and
Thomas Tooke

University of Toronto Press

©University of Toronto Press 1975
Toronto and Buffalo
Printed in Canada

ISBN 0-8020-2090-9 (cloth)
ISBN 0-8020-6214-8 (paper)

LC 73-84434

This book is dedicated to the preservation of the Niagara Escarpment

Contents

Foreword

Three hundred million years ago the area where the Niagara Escarpment now stands lay in sedimentary layers beneath a shallow sea. The area uplifted and warped, forming the saucer-like Michigan Basin. For millions of years afterward an ever-changing pattern of pre-glacial rivers and streams carved the Niagara Escarpment from the flat limestone plain. When, about a million years ago, glaciers began to advance and retreat over North America, they eroded away the soft under-layers of the escarpment, leaving a hard dolomite limestone cap at the top hanging high above. The result was a rolling gentle slope on one side and on the other a steep escarpment, slashed by gaps, padded with glacial debris, and riddled with streams which tumble over its cliffs.

Today the escarpment rises out of New York State south of Rochester and runs parallel to the shore of Lake Ontario all the way to Hamilton, whence it winds in snake-like fashion northward to the Blue Mountains at Collingwood. From there it travels west to Owen Sound and northward along the east shore of the Bruce Peninsula on Georgian Bay, dips under the water until it becomes Manitoulin Island and the continuous arch of islands westward, disappears again to reappear on the western side of Lake Michigan, and finally peters out in Wisconsin.

From the frontier at Niagara Falls to the tip of the Bruce Peninsula at Tobermory, the escarpment can be divided into three distinct parts: the north, Tobermory to Owen Sound; the middle, Meaford to Dundas; and the south, Hamilton to Niagara Falls. The first, once a fishing and lumbering centre, was settled last and is the most primitive; it now serves as a haven for tourists and summer residents. The second thrived in the early water-powered days of Ontario industry, and now is largely agricultural. The last, rich in the early history of Canada, is heavily settled and industrialized. The upper reaches of the escarpment appear much as they were in pioneer days, even though lumbering and fire have ravaged much of their natural beauty, many old buildings have been left to collapse or burn, and the tourist industry is taking its toll. At the southern end houses cling to cliffs, and smokestacks and high-rise buildings obscure the escarpment. All along the way, there is many a strange absence; mills which once dotted the escarpment mysteriously burned when their desperate owners could not forestall the onslaught of steam and electric power, churches have fallen into disuse, and houses have been abandoned. The story of many of the settlements is one of industry and commerce and then of people leaving, and of gravel pits and tourists and sportsmen proliferating. Where capitalism has thrived quite often the escarpment has not.

The twists and turns of the escarpment have had a lasting effect on our history from the earliest years. Tribal migrations and wars were directed by the lay of the land. Waterfalls, gaps, and portages marked the trails of the early explorers and the locations of early settlements. When we follow the route of the escarpment we follow the paths of war parties, of the confrontation of nations, of the founding of towns and governments, and of the growth of power, wealth, industry, and political movements. The escarpment is most of all a path of rapid change.

We have just begun to save the escarpment from decay and commercial exploitation. In 1960 a Hamilton resident named Ray Lowes, a lover of the escarpment and of hiking, convinced the Federation of Ontario Naturalists that 'No Trespassing' signs could be beaten, and soon the Bruce Trail was born. Today 450 miles of marked trail extend uninterruptedly along the top of the escarpment from Niagara Falls to Tobermory. More recently the government of Ontario has provided that a 'green belt' which includes most of the central part of the escarpment shall be reserved for farming and public recreation.

This book tells the story of the Niagara Escarpment, known as 'the mountain' to the pioneers and to local residents today. The photographs attempt to capture the unique beauty and neglected history of the region, and together with the text form a memoir of that which is rapidly slipping away.

ACKNOWLEDGMENTS

Our many thanks must go to a large number of people who helped in the preparation of this book, especially Mr Orrie Vail, Tobermory; Mrs M. Bogers, Lion's Head; the Reverend George Epoch, SJ, St Mary's Church, Cape Croker; the Reverend Thomas A. Scott, Wiarton; Mr C.C. Agnew, Industrial Commissioner of Owen Sound; Mr John J. Landen, Curator, The Grey-Bruce Museum, Owen Sound; Mr John W. Buzza, Leith; Mr Frank Harding, Meaford. Also Mrs W. Rutledge, Terra Cotta; Mr V.W. Vanderbrug, of the Hamilton Region Conservation Authority; Miss Olive Newcombe, Curator of the Dundas Museum; Mr Walter Reeves, Burlington; Mr and Mrs Beverly Flarity, St Catharines; Mr George F. Lewis, Winona; Miss Florence Martin, Curator of the Stone Shop Museum, Grimsby; and Mrs M. Rutherford, Grimsby.

The excerpt from Gordon F. Hepburn's description of the Bruce Peninsula bush fire of 1908 is reprinted from the Yearbook of the Bruce County Historical Society. William Green's account of his part in the Battle of Stoney Creek is taken from an article in the Hamilton *Spectator* of March 12, 1938. Edith Fowke has kindly agreed to the publication of verses from two folk songs, 'The Railroad Boy', printed in *Canada's Story in Song* by Edith Fowke and Alan Mills (Gage Educational Publishing Limited), and from 'The Wreck of the *Asia*' collected by Edith Fowke from C.H J. Snider and recorded in *Songs of the Great Lakes* by Folkway Records.

Publication of this book has been assisted by a grant from the Ontario Arts Council.

The Niagara Escarpment

Part one The North Section

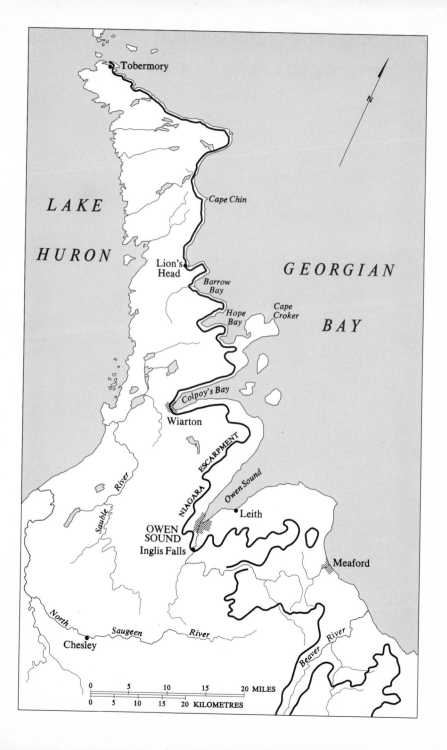

Part one The North Section
Tobermory to Owen Sound

Rugged cliffs looming over the shore south of Barrow Bay are typical of the harshly beautiful stretch of the escarpment that hugs the east coast of the Bruce Peninsula.

Behind the mist the blear sun rose and set,
  At night the moon would nestle in a cloud;
The fisherman, a ghost, did cast his net;
The lake its shores forgot to chafe and fret,
  And hushed its caverns loud.
. . .

When one strange night the sun like blood went down,
  Flooding the heavens in a ruddy hue;
Red grew the lake, the sere fields parched and brown,
Red grew the marshes where the creeks stole down,
  But never a wind-breath blew.

from 'How One Winter Came in the Lake Region,'
by William Wilfred Campbell (1858-1918)

The many bays and inlets along the eastern shore of the Bruce Peninsula, each surrounded by the cliffs of the Niagara Escarpment, are simply different entrances to the same stage. Everyone who has seen them has his favourite, to him surpassing all the others in beauty. But this peninsula was originally a massive battleground – scene of annihilative wars between the Hurons and the Iroquois, which raged there for years before the white man came. Its high cliffs and dense forests suited ideally their methods of combat. Later plentiful game lured French fur traders into the forests, but permanent settlers never arrived until the latter part of the last century.

Tobermory, in St Edmund Township at the tip of the peninsula, is one of the youngest settlements. Indians and traders had long roamed this country where Georgian Bay and Lake Huron join, but the harsh, rocky land hindered settlers and discouraged farmers, and treacherous currents impeded exploration even by water for many decades. Slowly in the nineteenth century the roadless land was opened by men attracted by the green gold of the lumber business. Then millions of board feet of pine, maple, ash, elm, and birch fed the five large lumber mills at Tobermory, and St Edmund township led the twin counties of Bruce and Grey in lumber production.

All over the peninsula, people moved in and lumber moved out. Great stands of forests fell until in some areas there remained only bare rock. In time the settlers were forced to turn to fishing. Tobermory's fishing fleet, over forty boats strong, braved the treacherous waters for many years. Some never returned. The two harbours of Tobermory, Big and Little Tub, have seen many ships slide in and out in the last hundred years. Both still have the aura of a port about them, for two car-ferries leave for Manitoulin Island several times a day. Watching and hearing them from the shore at night as they dock, it is easy to

Every year thousands of cars make the two-and-one-half-hour journey from Tobermory to Manitoulin Island aboard one of two ferry boats. The old dock has required expansion to accommodate the ever-increasing tourist trade.

The neatly trimmed harbour at Tobermory, lined with the yachts of cottagers and visitors, seems to have banished the curse of death and destruction which once plagued sailing ships in the moody waters nearby.

imagine that they are laden with furs or lumber, or carry intrepid explorers to destinations unknown. The pleasure craft lining the harbours belie the fact that many lives have been swallowed up nearby, and that the lakebed is littered with wrecks.

In 1872, the steamer *Mary Ward*, after travelling all the way from Sarnia to Owen Sound, foundered on a shoal on a calm evening during the last leg of the trip to Collingwood. In 1906 the steamer *Jones*, on its regular run from Owen Sound to Tobermory, simply disappeared after passing Cape Croker, taking passengers and crew with her. During the Great Storm of November 9, 1913, over twenty ships and 250 men were lost, with a million dollars worth of cargo. While steamers sank into the deep on that black day, all of them huge steel freighters, one small wooden three-masted schooner, the *Sephie*, survived unscathed in Georgian Bay. Its captain later commented: 'At one time they made ships of wood and men of steel, now they make ships of steel and men of wood.'

The most widely remembered tragic ballad of the Great Lakes sings of the wreck of the paddlewheeler *Asia* in 1882, with a loss of more than 200 passengers and crew. She sank in broad daylight on a run from Owen Sound to Manitoulin Island, leaving only two survivors to wash ashore near Parry Sound. 'The Wreck of the *Asia*' was sung for years all around the Georgian Bay area, with the final melancholy verse:

Now in the deep their bodies sleep, their earthly trials o'er.
And on the beach their bones do bleach along the Georgian shore.
Around each family circle how sad the news to hear,
The foundering of the *Asia* left sounding in each ear.

Perhaps the most famous shipwreck within sight of the escarpment was that of the *Griffon*, built under the orders of René-Robert Cavalier de La Salle, the French explorer who founded Lachine on the Saint Lawrence and whose travels inland carried him to the mouth of the Mississippi. A sixty-foot ship made of oak and carrying five guns, the *Griffon* was built 'seven leagues above Niagara Falls' in 1678 to extend the fur trade northward where no ship had sailed before. Her name came from the mythical creature, part lion and part eagle, which was emblazoned on the coat of arms of La Salle's patron, Count Frontenac, Governor of Canada. Her maiden voyage – the first by any sailing ship on the Great Lakes – carried the *Griffon* across Lake Erie, up the Detroit River into Lake St Clair, and across Lake Huron and Lake Michigan. She took on twelve hundred pounds of furs, valued at sixty thousand francs, and thus laden sailed back without La Salle, under the command of his pilot. On September 18, 1679, she sank with all hands aboard. Her whereabouts remained a mystery for almost three centuries. Then, in 1955, Orrie Vail, a Tobermory fisherman-patriarch, found the remains of the ship on the lake bottom in a small cove off

Orrie Vail's museum of nautical relics at Tobermory is built mostly of timbers salvaged from shipwrecks. Visitors peer through chicken wire lining the narrow aisles and crammed shelves of the museum. One of Mr Vail's most prized exhibits is a letter authenticating his discovery of the Griffon.

Orrie Vail is famous for his hand-made knives, produced at the museum while he entertains his guests with tales from local history.

Four miles offshore from Tobermory lies Flower Pot Island. The two limestone-topped formations have been slowly sculpted by water erosion, which has eaten away the softer shales at the base. The larger flower pot has been shored up by man. It is about fifty feet in height; the other is twenty feet shorter. When explorers first discovered the flower pots there were three of them. The 300-acre island, riddled with caves and cliffs, now is a national park.

Russell Island at the tip of the peninsula.

Some people say that Orrie Vail *is* Tobermory. He presides over a shingled storehouse of local history, a museum of relics, fossils, and salvage from old ships which sailed the local waters in the days of his father and grandfather, an original settler of the region. The building was constructed in 1889 of lumber salvaged from shipwrecks. Orrie tells many fascinating stories and legends, one of which is the Indian tale of the nearby Flower Pot Islands. This version is taken from *Tall Tales and Collections of the Georgian Bay* by Mrs Melba Croft.

Many moons ago, when the Indian people owned all the land from the Manitoulin to the Blue Mountains, tribes lived within a few arrows' flight of one another. There were many wars. Sometimes young braves of one tribe fell in love with young maidens of another. This was not good, for each tribe was very proud and independent. Toward the end of the eighteenth century, the son of a powerful tribal chief fell in love with the daughter of another chief, and ran away with her.

The lovers travelled by canoe. Soon the maiden's father assembled his warriors and set off after them. Now as he paddled swiftly, the young lover remembered the Island of the Caves and hurried there to hide his sweetheart in a deep cave. But the maiden's father had also thought of this and soon caught up with the pair. The young brave was killed, and the maiden died of a broken heart.

Since that time, Indians have called the island the Island of the Flower Pots, and they claim that the two flower-pot formations are the stone spirits of the young Indian lovers. To them that island is a forbidden place, and they will not set foot on its shore.

The famous Bruce Trail begins at Tobermory and runs south along the escarpment all the way to Niagara Falls. In this part of it hikers climb over rocks and through forests along the many bays of the peninsula.

Each of the peninsula's beach-hugging settlements bears the mark of a history of struggle and achievement. Cape Chin boasts the Church of St Margaret, which was built over a seven-year period by the townsfolk out of local wood and escarpment stone, quarried within a mile of the site. The Reverend Canon R.W. James served in its construction as lumberjack, hewer, sawyer, quarry-man, stone mason, and building superintendent. His Model T Ford served as transport truck, and even as power plant, while he rounded up local volunteers to haul two-ton blocks of stone (upside down because the bottom, which had never been exposed to air, was soft) and to build the structure with them. Contributions came from passers-by; visiting anglers from the city not only threw in a large donation (rumoured to be the proceeds of a poker game) but also inspired the folks back home to contribute windows and pews. After the mammoth community effort, Archbishop Charles Seager

This restored pioneer log cabin, built in the bush around 1850, is located at the site of the Tobermory Museum.

The stone Church of St Margaret at Cape Chin, built entirely of local materials by the congregation, is today a little-used testimonial to the community spirit of those who once lived here.

opened the church in 1932. Services were held every second Sunday before about twenty-five people, but attendance has dwindled in recent years as people have moved away, and now the church which was once the pride of the community houses services only two or three times during the summer season. The Hayes family, who donated the land, still farm around it.

Perhaps the most distinctive of the Bruce's bays is at Lion's Head. The natural rock formation which gives the site its name and guards its entrance has sheltered boats for years. Much of the original rock has crumbled into the bay, but jagged remains, about 168 feet high, still jut from the promontory known as Lion's Head Point, a sailor's landmark for passing Great Lakes vessels since the days of sailing ships. The headlands form a ridge of limestone known as Hamilton Mountain. Along the shore near the head is a cave-like rock formation known as Eagle's Nest, where a small boat can take shelter. A natural stone bridge stands on the Bruce Trail just north of the town.

The first settlers in this area, arriving in the late 1800s by land amid dense bush and high cliffs, did not realize that a harbour lay only two miles away. Later the area became a bustling trading and fishing centre traversed most easily by water; the steeple of the Anglican church guided many a local fishermen into the harbour. Telegraph service arrived in 1887, but without easy land routes communications remained arduous. One day a message arrived for David Porter, a candidate for the provincial legislature who was campaigning on the peninsula. One of the settlers, Donald McIvor, walked fourteen miles to deliver it – one-third of the way over a path that snaked through dense bush broken by only a few small cleared patches. The telegram informed Porter that five hundred dollars had just been voted for road improvement. McIvor got the then-princely sum of two dollars for his labour.

The lumber industry was still booming in some parts of the peninsula as the century drew to a close. Telegraph poles for the United States, logs for the saw mills at Wiarton and Owen Sound, and ties for the Grand Trunk Railway flowed from local forests. As the trees dropped, settlers came in droves. Not all by any means were literate – nor even some of the tradesmen who supplied them with food and tools. A story is told of one genial but unlettered shopkeeper who made up for his deficiency with art. If goods were not paid for, he drew a picture of his customer and the article in question. One day he told a man that he still owed the price of a big wheel of cheese. The man claimed that he had never bought more than fifty cents worth of cheese at once in his entire life, but the trader produced his book with a drawing as proof. The man pondered a bit. Soon he erupted in laughter. 'I bought a grindstone from you last fall – you forgot to draw the hole.'

Barrow Bay, a summer community a few miles south of Lion's Head, is the location of the Greig Caves, which penetrate the lower levels of the escarpment. Visitors climb down to them through a three-foot-wide crevice, clinging to a steel cable.

At Lion's Head cottagers have replaced fishermen in the shelter of the bay.

The Greig Caves at Barrow Bay, formed by the pounding of Georgian Bay waters, are marked by a massive rock overhang at the entrance.

The undulations of the escarpment's headlands fade into the distance looking north from Hope Bay.

The almost two-hundred-year-old white board Jesuit church on the Cape Croker Indian Reserve served as a school until recently and is still in use as a meeting hall.

Beyond the harsh, rugged terrain of Cape Dundas lies Hope Bay, another tourist mecca. Once a thriving lumbering community whose speciality was tamarack bark, used in tanning, it now is famous for its sandy beach and beautiful scenery. The shores are lined with sheer limestone cliffs, with huge boulders at their base near the water's edge. Atop the cliffs are the Indian Wells, potholes three to ten feet wide along the Bruce Trail, formed by glacial waterfalls. Their depth is unknown.

The Cape Croker Indian Reserve, home of Ojibway Indians, includes 550 acres of camping park as well as some famous scenic caves. Here the Bruce Trail passes through primitive dense forest. In the mid-nineteenth century, a local Indian princess called Nah-nee-bah-wee-quay, who was married to a pioneer, William Sutton, travelled to England to plead for the preservation of her lands and her people. Granted an audience by Queen Victoria, she was treated as a royal equal, listened to sympathetically, and sent home with royal gifts. But the plea was lost in bureaucratic red tape and Nah-nee-bah-wee-quay's people were forced on to Cape Croker Reserve, just as their more southerly fellows had been forced on to reserves in decades past.

The reserve itself has long been a centre of Christian worship. According to the terms of an 1857 treaty, the Newash, a branch of the Ojibway nation, surrendered their lands and moved to the reserve, but requested that the Jesuit Fathers continue to minister to them. A Father Falhuber arrived shortly afterwards to erect a log church, but for nearly half a century the reserve had to make do with only occasional visits, three or four times a year, of priests who travelled from their headquarters on Manitoulin Island to visit the surrounding communities. Then, in 1902, the Jesuit Provincial sent a resident priest, and in 1904 came Father Cadot, son of a wealthy Montreal family, who stayed for twenty-seven years. The tribe named him Wallasseshkong, 'Brightest Light to Shine.' The present incumbent is Father Epoch.

Father Falhuber's church burned in 1859 and was replaced by a white board-batten building in the same year. The new church served until St Mary's, built solely by Indian labour, opened in 1908. The old wooden church served as a school until recently, when it became the community meeting place and parish hall.

The United Church also has a house of worship on the reserve. It was erected by contractors from Chesley who never built again. They drowned in the bay during the trip home. Their Indian guide, who had warned them not to go, skilfully survived the ordeal but left the boat to rot on the beach, for such vessels he knew were cursed by evil spirits.

This guide was typical of the local Indians, who lived in close communion with the waters of Georgian Bay. So too is the story an old chief used to tell of one of his great grandfathers, also a chief, whose people once were ice-bound at an early spring sugar camp seven miles across Colpoy's Bay from their settlement. The chief made a small cedar paddle on which he drew a sun with charcoal; then he laid the paddle on a rock, and prayed to the Great Spirit to help him to save his stranded people. Suddenly a crack opened in the ice straight to their home. As soon as they touched the opposite shore, the legend has it, the miraculous crack disappeared.

Any history of this part of Ontario would be incomplete without mention of the great bush fire of 1908, which burned over most of the Bruce Peninsula north of Cape Croker. This terrible conflagration started near the Lake Huron shore and, fed by the heaps of dried and rotting brush that had accumulated in the woods over many years of lumbering operation, did not stop until it reached Georgian Bay. Gordon Hepburn, who was a five-year-old at the time, living in Hope Bay, has described the scene:

At about ten o'clock in the forenoon we noticed a pillar of smoke rising up, increasing by the minute, and quite quickly rolling out over the whole of the western horizon.... The men rushed to remove portions of log, rail and brush fences, which all too often were built right up to the corner of a building, to get all animals and persons out of buildings, to ready containers of water...

At about eleven o'clock the smoke began to reach us, very quickly thickening around us and soon blotting out the sky and the bush outline.... The sun began to fade into a copper disk...it seemed that even the sun was deserting us. Our home and the Bruce Peninsula was then our whole world, and it was being swallowed up and engulfed in a blinding pall of acrid smoke.

As time went on, the heat from this wall of fire began to reach us. My mother put wetted cloths over our faces and had us lie flat on the ground in the yard. We were hemmed into a visibility of a few feet now; we could hear timber crashing, occasionally a frenzied animal rushing madly by in the smoke, a horse pounding off to the side, plaintive bleat of a sheep or lamb, a cow bell's staccato jangle. ...

As the heat and the smoke increased until it was almost unbearable, with embers falling around, firing our clothes at times, and the ground covered with ash as a light snowfall, we began to have difficulty in breathing. Our poor eyes were nearly burning out of our face.

Very shortly after this the fire crowned over us, leaping forward as if in an explosion, to travel over the clearing and catch on bush beyond to the east of us now, racing to build up in the sky line toward the Georgian Bay.

It was five years before the settlers could begin to think about farming and building their lives again.

Wiarton, 'The Gateway of the Bruce Peninsula,' was the site of Indian mission stations during the early seventeenth century. First settled in 1866, the harbour on Colpoy's Bay once served a small fishing fleet; today it provides anchorage for pleasure craft.

The 450-mile-long Bruce Trail winds along this road on the Cape Croker Reserve.

Scenes such as this in Wiarton harbour are a far cry from the view in W.W. Campbell's youth, one hundred years ago.

Centennial Tower on East Hill, overlooking Owen Sound, was built by the students of the community's two high schools. It provides a majestic view of the valley, which has been a village site since Indian days.

One of Wiarton's most colourful early characters was also one of Canada's earliest women's liberationists. About one hundred years ago Margaret Paterson was the only person in the area with any medical knowledge. As a girl in Scotland, she had had the single desire to be a doctor, but medical schools then refused to admit women. As a second-best, she armed herself with a crash course from a doctor brother, then married an apothecary and in 1869 set off for Canada. She made her rounds in the Wiarton area in a horse and buggy, tending the sick, for many years.

The bard of the Great Lakes, William Wilfred Campbell (1858-1918) spent many years of his youth in the area, primarily in Meaford and Wiarton, where he fell in love with the peninsula and the blue waters of Colpoy's Bay, and began to write his Lake Lyrics. A popular poet in his time, he returned to his favourite theme again and again over the years. A cairn has been erected as a memorial to him in Wiarton Park.

Today on the Bruce Peninsula, settlers' cabins have been replaced by concrete and steel; paved roads run where once only forest trails existed; once-prosperous industries have disappeared. Only the solidity of the escarpment remains unchanged, and to this permanence thousands of tourists flock each year.

At the peninsula's base, the escarpment is known as 'The Hill.' Here the city of Owen Sound sprawls between two large bluffs in a region called by the Indians Wadineednon, or 'Beautiful Valley.' The first white men arrived here in 1616 led by Samuel de Champlain, and found a huge Indian encampment already on the site. One hundred and seventy-two years later another traveller, Gother Mann, took shelter in the bay during a severe storm and he too saw Indian villages where Owen Sound North now stands. The wars of the Iroquois and Huron nations never quite reached the Owen Sound Valley but several large and bloody battles took place in the surrounding hills, and one huge burial mound still rises west of the city on the Saugeen River.

Historical documents say nothing of the Owen Sound Valley until 1815, when the noted cartographer, Captain Fitzwilliam Owen, arrived in the bay on a mapping expedition. The land then was still rich with the untouched raw material of the fur trade. Twenty-two years later Charles Rankin, the provincial land surveyor, started laying out the area. By 1840 John Telfer began building a log complex in the Market Square area, including three buildings known as Government House.

When the government offered each man fifty acres of land, settlers hurried in. The town's location was ideal for the developing trade with the western reaches of Canada and the United States. It was then called Sydenham, but was incorporated as the town of Owen Sound in 1857. The natural harbour soon became one of the town's main resources. The first steamer in the bay was launched in 1867 and many wooden boats were built on its west shore.

In 1872 the Toronto, Grey, and Bruce Railway arrived, and soon folks were

The grain elevator on the dock at Owen Sound.

Tom Thomson, probably the most famous Canadian painter, grew up in this house in Leith, to which he moved with his parents when he was two years old.

The body of Tom Thomson rests beneath this headstone in the churchyard at Leith. Although there has been some dispute over whether the body rescued from Canoe Lake was in fact the painter's, relatives who attended his funeral have verified the fact.

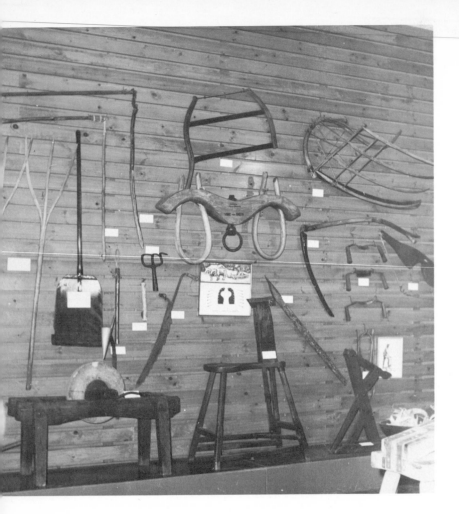

Early pioneer implements on display in the Owen Sound Museum.

altering an old Irish song by singing:

> If I had all the riches that's in my father's store
> Oh freely would I share it with the boy that I adore.
> We'll fill our glasses to the brim, let the toast go merrily around,
> And we'll drink the health of the railroad boy, from Ottawa to Owen Sound.

The railway turned Owen Sound into a bustling metropolis. The Canadian Pacific Railway took over the line twelve years later and built a wooden grain elevator on the east side of the harbour, and ships like the *Manitoba* and the *Athabaska* carried freight to Port Arthur (now Thunder Bay) for shipment to the Prairies. But in 1911 the elevator burned down, and the CPR consolidated its operations at Port McNicholl, near present-day Midland at the southeastern corner of the bay, which was closer by rail to Toronto. Nevertheless Owen Sound continued to grow and by 1920 was incorporated as a city. The Great Lakes Elevator Company built a million-bushel grain elevator there in 1925, and later enlarged the capacity to four million bushels.

Tom Thomson (1877-1917), perhaps the most famous of all Canadian painters, grew up in Leith, six miles from Owen Sound. Thomson traversed much of northern Ontario during the last three and a half years of his life, during which he devoted all his time to painting. By that time he was already represented in many public and private collections and was firmly established as leader of a new movement in Canadian art. His death by drowning on Canoe Lake in Algonquin Park is a mystery, for he was an expert canoeist and the day was calm: his body rests in the graveyard at Leith. The Tom Thomson Memorial Gallery and Museum of Fine Art in Owen Sound displays his paintings, and works of his friends and associates, Jackson, Varley, Casson, Carmichael, and Lismer.

Other paintings are housed in the County of Grey Owen Sound Museum on the east hill of the city. This museum portrays the natural and social history of the county and exhibits records of early settlement along with collections of artifacts, tools, art, and pioneer building. Its main log house was built in the 1830s west of Meaford by William Raven; it has been completely restored and has been joined by a smaller cabin and a blacksmith shop.

Behind the Harrison Park Complex at Inglis Falls, the Sydenham River drops about 100 feet over the escarpment on its way to Georgian Bay. This famous falls, a major point of interest on the Bruce Trail, provided power for the first grist mill in the area and later the first electric power. The original millstones are on display above the falls.

The white man touched this northern end of the escarpment last and least. As it plunges southward the escarpment becomes richer in his history, his diversity, his frailty, and his folly.

The dam on the Sydenham River above Inglis Falls once stored water for a long-since disappeared mill.

The churning 100-foot drop at Inglis Falls, where the Sydenham River flows over the escarpment at Owen Sound.

One of the old millstones at Inglis Falls.

This natural stone bridge graces the Bruce Trail on the escarpment above Lion's Head.

Part two   The Middle Section

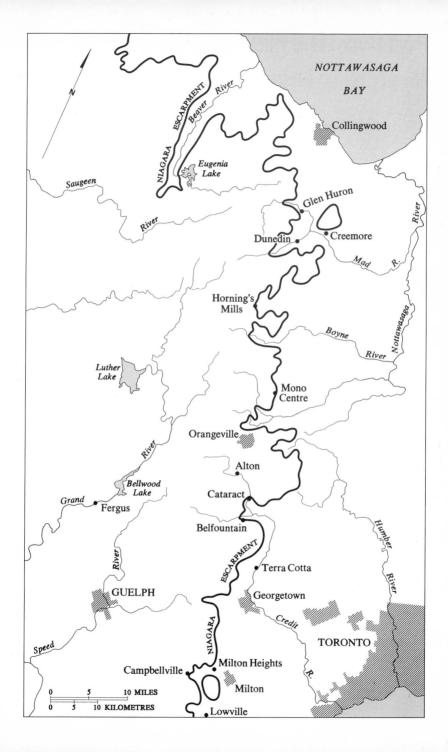

Part two The Middle Section
Meaford to Dundas

The lush farm land beneath Mount Nemo typifies the pastoral scenery along the middle section of the escarpment.

originally thickly wooded to their summits, but now seamed with roads and interspersed with clearings. Probably none of us would have noticed them, though their beauty is enough to attract passing attention, had they not been pointed out as the highest 'Mountains' in the great Province of Ontario!
from *Ocean to Ocean, Sandford Fleming's Expedition through Canada in 1872* by George M. Grant, secretary to the expedition

The modern technological era has not treated the small settlements of the escarpment kindly. Their days of glory have ended, along with the water-driven industries which nourished them. But when progress passed them by it left a wealth of beauty and history. Beyond Owen Sound, the escarpment runs around the south shore of Georgian Bay, through small places with names like Paynter Bay, Coffin Cove, Johnson Harbour, and Cape Rich. Near Meaford, an army tank range covers a large area that is inaccessible to the public, but then the escarpment is open again, running south along the Beaver River to Eugenia Falls and back to the bay at Collingwood. From there it travels almost due south, all the way to Burlington and Dundas on Lake Ontario. The towns in this section provide a glimpse of the Canada of over one hundred years ago. Their stories centre around hardy settlers and shipping, flour mills, saw mills, grist mills, and woollen mills; around booming prosperity followed by the accidents of history and the inventions of men, which led to gradual decline and ultimately the drifting away of population. Today the region is principally a haven for tourists and sportsmen, and lovers of the smell of pine and the rush of waterfalls.

The first known settler in the Meaford area was William Corley, who arrived from Ireland in 1830. He was already established when the first survey of the area was carried out—in fact, his sons assisted in the operation. Just who was first to live at the actual mouth of the Big Head River was a matter of dispute. Two men, Londry and Whitelaw, arrived there at about the same time early in 1834; Londry claimed the honour, however, because his was the first cabin completed. About eight years later Donald Miller built a shanty near the Big Head River, close to the centre of the present town. Most of the subsequent settlers arrived by boat at Peggy's Landing, named after Miller's wife.

Soon after he was settled, Miller harnessed the river for a grist mill. In 1845 the town was named Meaford in honour of the county seat in England of the Earl of St Vincent, a British naval hero of the American Revolution and Napoleonic Wars. By 1905 it boasted about thirteen industries, but none was more important than the harbour, where ships called to transship goods to western Canada. The railroad arrived in 1872, but was built on top of the mountain far from the harbour. Today the escarpment-surrounded harbour is a haven for small pleasure craft; and the Meaford Museum, housed in the old water works building, displays early farm implements and pioneer furnishings.

At Meaford the escarpment surrounds a harbour for the last time before plunging southward along the Beaver River.

*From the Bruce Trail, the east side of the escarpment can be seen looming over
the Beaver Valley.*

At Eugenia Falls, the Beaver River plunges over the escarpment from Eugenia Lake. Here the escarpment turns northward again, along the river's east side.

South of Meaford the Beaver River Valley runs southwestward for about thirty miles to Eugenia, through one of the most scenic regions on the escarpment. Eugenia Falls, at the lower end of the valley, has a seventy-foot drop. It was unknown until 1853, when a man named Brownlee, one of the earliest settlers of the area, heard its distant roar while hunting, and followed the sound through thick unexplored forest for five miles. When he saw the falls he was spellbound, and rushed home to tell the nearest neighbour. The two men returned, climbed down the walls for a better look – and at the bottom saw a yellowish metallic substance gleaming through the spray. They decided that it was gold, and swore each other to secrecy.

But Brownlee still couldn't keep the story of the beautiful falls to himself. Soon curious townspeople were at the site and, peering over the edge of the gorge, saw the two men hiding something at the bottom. The local gold rush was on. It ended only when one thorough-minded adventurer sent a sample to be assayed and learned it was worthless pyrite. Only one man made a profit from the fiasco at Eugenia: he sold a sack of the 'fool's gold' to a gullible farmer who had not yet heard the dénouement of the discovery.

The gold rush did bring Eugenia Falls to the nation's attention. A town was laid out and named after the French Empress Eugénie, wife of the ill-fated Napoleon III; the streets were named for the great battles of the Crimean War. Several dam sites were built above the falls and the town prospered for a while, but industry gradually disappeared and the population drifted away.

One of the inhabitants, however, had a dream. William Hogg built at Eugenia a small electric plant consisting of a timber crib, a dam, a timber flume, and a powerhouse. Two forty-horsepower turbines drove a generator which provided power for Eugenia and Flesherton. In 1915, the Ontario Hydro Electric Commission opened at the same falls a greatly enlarged plant with a rated capacity of 4,500 horsepower. The plant cost about $1,190,000 and was the second built by the commission while Sir Adam Beck was its chairman.

From Eugenia Falls the escarpment runs up the east side of the Beaver Valley and eastward to Collingwood, nestled in the shadow of the Blue Mountains and Osler Bluffs on Nottawasaga Bay. Here the escarpment reaches its greatest heights, 1,100 feet above Georgian Bay. Two hundred years ago the region was the home of the Tobacco Indians. Although the first settler arrived in 1835, and Hurontario Mills (now called the Old Village) was founded in 1840, Collingwood did not grow until in 1852 its location was chosen as the northern terminus of a new railroad. The next year, industry began with the construction of a sawmill. The town had originally been called Hens and Chickens, in reference to a cluster of rock islets offshore, but with its new importance it was incorporated and renamed after a British admiral who was Nelson's second-in-command at Trafalgar. In 1872, the secretary of an expedition headed by Sir Sandford Fleming (who was then serving as engineer-in-chief of the survey for

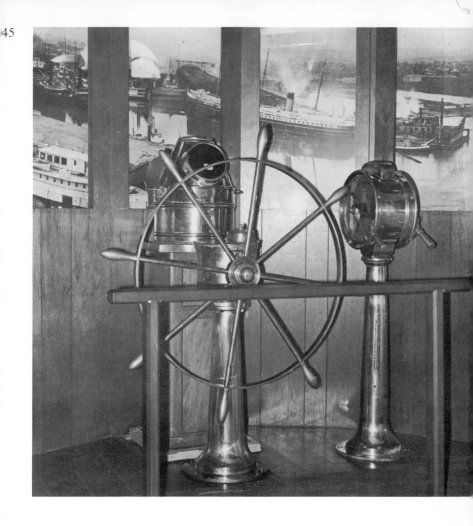

This steering wheel and telegraph in the Collingwood Museum were taken from the s.s. Southton, *built in England in 1928 and later renamed the* Gray Beaver.

Collingwood is an instance of what a railway terminus does for a place. Nineteen years ago, before the Northern Railway was built, an unbroken forest occupied its site, and the red deer came down through the woods to drink at the shore. Now, there is a thriving town of two or three thousand people, with steam saw-mills, and huge rafts from the North that almost fill up its little harbour, with a grain elevator which lifts out of steam barges the corn from Chicago, weighs it, and pours it into railway freight waggons to be hurried down to Toronto and there turned to bread or whiskey....Around the town the country is being opened up, and the forest is giving way to pasture and corn-fields.

Shipyards have been a part of the community's life blood from its earliest days; looking down the main street today, shipyards still fill the view. The harbour was opened up in the 1850s and by the turn of the century three large shipping firms were doing a roaring trade. More recently, however, trucking companies have cut heavily into the rail and shipping monopoly, and today Collingwood is famous as a tourist centre rather than an industrial one. Its attractions include the Blue Mountain ski runs, the weirdly beautiful scenic caves, the Collingwood Museum, with its wide variety of Indian relics and shipping memorabilia, the Blue Mountain Pottery, famous for its ceramics, where visitors can watch master potters at work, and the Huron Institute with its relics of Indian days.

The escarpment continues southward through the small settlements of Glen Huron and Creemore on the Mad River. Creemore, first settled in the 1840s, was typical of the small towns which grew up as a result of the escarpment's cheap water power. Quickly it accrued a saw mill, a grist mill, a woollen mill, two telegraph offices, a machine shop, and even the county's first policeman.

From here the escarpment passes through beautiful scenery to Dunedin and to Horning's Mills. The latter centre was founded in 1830 by a settler from the Hamilton area, Lewis Horning. When he led his expedition of hardy and hopeful families over rough terrain into this previously untouched region, he had great plans for industrial exploitation, wealth, and the establishment of a large and bustling community. But tragedy put an end to all that when four children, one of them his own son, overheard him offering a dollar to the man who would scour the surrounding woods for a stray cow and calf. Determined to win the prize themselves, the children disappeared into the forest. They were never seen again. A search party followed their tracks until they met those of two Indians. Many years' searching turned up only one of the boys, and his faulty memory did little to clear the mystery. He did say, though, that the youngsters had been taken away by the Indians and hidden from white men's

*The working floor at the Blue Mountain Pottery can be seen from a visitors'
observation area. The local clay is finished with the distinctive blue-green
glazes for which the pottery is famous.*

The wide main street of Orangeville is typical of Ontario's rural community centres, where farmers for miles around come to shop, do business, and socialize.

Built around 1850, the oldest remaining stone house in Orangeville is now an apartment building.

Thousands of skiers flock every year to modern runs, such as this one at Hockley Hills, with their scenic settings, modern accommodation, and club-houses.

view at all times. Many of the settlers were frightened away by the incident and even Horning himself left, broken-hearted at his loss.

Passing through Mono Centre, the escarpment leads next to the Dufferin County seat, Orangeville, near the headwaters of the Credit River directly west of the Hockley Hills. The surrounding area is the source of five of the river systems of southwestern Ontario: the Humber, Grand, Credit, Nottawasaga, and Saugeen. This abundance of water and water routes played a major role in the early development of the district, which had been a favourite hunting ground of the Mississauga Indians.

The first grant of land near Orangeville was given in 1820 to a surveyor, Ezekiel Benson, who was the sole landowner until 1822. Gradually a small settlement developed, at first simply named 'The Mills' because of the saw mill and flour mill built by James Greggs in 1832. Orange Lawrence arrived in 1843. Born in the state of Connecticut in 1796, Lawrence had moved north to Canada with his family and had first settled in Halton County. He later bought land in the Orangeville area and built a saw mill which spawned a town. He died in 1861. Today, the community named after him is a substantial farming and industrial centre.

In the surrounding countryside, a burgeoning winter sports industry has developed – as it has in many other areas along the escarpment. Skiing has become common, and chalets, motels, and lodges have sprung up to accommodate the winter sportsman and his family. In the undulating hills around Orangeville there are excellent ski runs at such resorts as Hockley Hills, Mono Mills, Twin Hearths, and Valley Schuss. Close to the Forks of the Credit, the Caledon Ski Club offers several good runs on fast hills with modern up-hill towing equipment. Further north, there are other exciting runs and comfortable shelter in the Blue Mountain and Beaver Valley areas. Further winter sports centres can be found dotted along the length of the escarpment from Hamilton to Owen Sound.

Along the Credit River valley and through the Caledon Hills the escarpment next runs. Here precipitous heights enclosing the sinister-looking, vine-hung Devil's Pulpit slowly give way to flat land surrounding the settlement of Terra Cotta.

The Credit River has entrenched itself as deeply in the history of this area as it has in the rock of the escarpment. Long before the first white man dreamed of the New World and its riches, Indians canoed the Credit's waters. The French fur trade brought both prosperity and decline, as the Indian nations slowly withered from sickness, war, and the white man's greed. The river was too valuable an asset to lose: smooth millponds, massive water wheels, and rushing white water ushered in the escarpment's industrial age.

In 1834 Thomas Russell and his family settled the upper reaches of the river at Alton. For three long and lonely years, the family survived the freezing winters and vastness of the land alone. Eventually the lure of the wilderness

In Alton, McKenzie's Mill, formerly the Caledon Centre for the Arts and Crafts, is situated beside this dam on the Credit River.

All that remains of the old mill and race at Cataract.

brought other settlers to the unknown region, and by 1851 the community had a grist mill built by Messrs Shingley and Farr and a store operated by Robert Meek. In 1885 Alton was recognized officially with the granting of a post office. The kilns of Jamison and Carroll produced prized quality lime. By 1877, Alton was a thriving industrial and agricultural community. But water power was also Alton's ruination: a disastrous flood in 1889 ravaged much of the settlement and its industry, and the community never quite recovered.

For about a decade the Caledon Centre for the Arts and Crafts was located in the old Alton Knitting Mill. This structure, built in 1881, has survived both flood (1889) and fire (1917), and continued in operation as a mill until 1963. It is a handsome and silent testimonial to the vanished decades it has witnessed. Recently it changed hands and was renamed McKenzie's Mill.

The Credit River had its share of dreams as well as of the sawdust and millstones of the industrial age. In 1818, in Cataract, or Churches' Falls as it is commonly known, the cry of 'gold' was briefly heard. 'There's gold in the hills behind the town!' Settlers deserted the farms, people poured into the valley, and expectations of sudden wealth ran high. But there was no gold; the would-be miners drifted back home.

Yet not all was lost. During the gold rush, a young man by the name of Grant discovered a salty spring near Credit Falls. After a bit of cajoling, and a few years, salt miners began to exploit the white gold beneath the ground. A saw mill was built to furnish lumber for the mine and for a splendid new town that was to be named Glennifor. But this too proved all a dream. No salt of any consequence was found, and the entire effort had to be abandoned.

Yet Cataract itself did not die. A woollen mill was built, and the town was chosen as a junction for a proposed Credit Valley Railway. During the second world war the woollen mill was converted into the Cataract Electric Company, which furnished desperately-needed power for the valley. But at the war's end the plant was closed and the millpond dam was partially blown up. Ruins are all that remain of a settlement which once had so many dreams.

South of Belfountain, the torrent and turbulence of the Credit River gives way to meandering ripples. The industries of the river's upper reaches surrender to rich farmland. Salmonville (Terra Cotta) sprang up along the old wagon road to Oakville on Lake Ontario. This was a busy route during the Crimean War, when a seemingly endless need for supplies kept horse-drawn wagons plodding back and forth between the neighbouring rich wheat farmlands and the Oakville wharves. Teamsters took full advantage of Salmonville's three inns and its blacksmith.

Once peace returned to the Crimea, great clay deposits brought other wealth to the settlement – the terra cotta of kilned bricks. Simon Plewes – owner of a large flour mill on the Credit banks – had discovered the clay years before and had constructed his house from hand-made bricks kilned in his own yard. As the

Before the turn of the century Simon Plewes built his home in Terra Cotta from bricks made of clay in his own yard, thus giving birth to the local brick-making industry which gave the town its name.

The Credit River exhibits many moods as it travels southward to Lake Ontario. Here it flows through a valley just south of Terra Cotta.

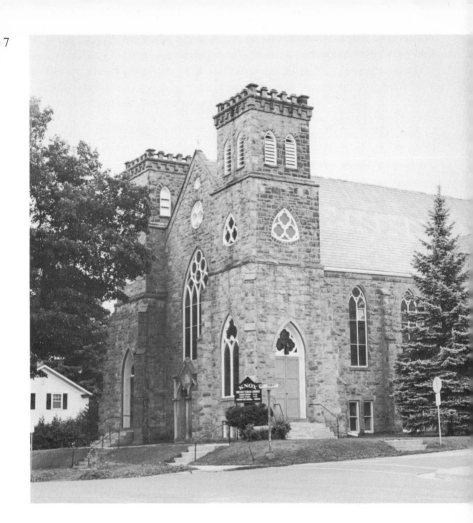

Georgetown's Knox Presbyterian Church, built in 1887, is typical of the Ontario architecture of that time.

ville officially changed its name to Terra Cotta.

Today Plewes' mill is in ruins; the brick yards have closed. Instead of wagons, gravel trucks roar along the roads near this sleepy community. The old forge, once the studio of sculptress Rebecca Sisler, still stands, however, as does Plewes' home and old inn.

In the lee of the escarpment, among the hills which surround the west branch of the Credit river, lies the picturesque town of Georgetown. The town is named for George Kennedy, the first settler, who arrived in 1823. He took up a 200-acre tract of land in the area then known as Hungry Hollow, and built a saw mill on a small stream that flowed across present-day Main Street. The mill's powerful overshot wheel took its strength from the high drop of the water-course. For some years there were only two other settlers, Marquis Goodenow and Sylvester Garrison, but in 1837 the four Barber brothers arrived in Georgetown and during the next thirty-nine years they acquired immense wealth in partnership, owning almost everything in common while they built a flourishing town.

The Barber family had come originally from Ireland to Crooks Hollow, near Greensville, where the boys' father had woollen, paper, saw, flour, and grist mills, a cooper shop, a distillery, and a tannery. Here the boys learned their trades: William and Robert in the woollen factory; James in the paper mill; Joseph in the foundry and the millwright's shop. When they struck out as men for the Credit River, their eyes were on the water power of its west branch, and with the thirteen acres they purchased from George Kennedy they also acquired water rights. One by one the mills went up. The first was a woollen and carding mill, established by William and Robert along where Water Street now lies. Joseph built a foundry, which prospered with the construction of the Grand Trunk Railway. James's paper mill – starting with rag-based products – grew in the twentieth century to one of the largest papermaking plants in the country before it was closed in the 1950s. Georgetown grew with its industry. In 1864 it was incorporated as a village and thirteen years later it had 2,500 residents. Its chief constable at the time was considered notable, not for any ferocity of appearance, but for an apposite name which, combined with shrewd detective ability, made him a terror to evil-doers: his name was Edwyn Search. Georgetown also had a volunteer fire company under the captaincy of John R. Barber. About 1887, a private company was formed to generate electricity at Glen William, and subsequently the town streets and houses were lighted for the first time by hydro power. The Aetna Lacrosse Club brought many honours to the town during the 1890s.

Further south three high cliffs, Milton Heights, Rattlesnake Point, and Mount Nemo, loom above rifts in the escarpment. The dramatic precipice of Milton Heights hangs close to Highway 401 as it passes through a gap in the escarpment just west of Milton. The Heights wind tortuously southward to the

On the Bruce Trail, approaching Milton Heights — a precipice visible for many miles around.

Approaching Mount Nemo from the west on the Bruce Trail.

steepest and highest of the slopes, Rattlesnake Point. Mount Nemo lies southeast of Lowville at the Bronte Creek gap, where the rock is penetrated by Mackenzie's Cave, so named because the rebel William Lyon Mackenzie is said to have hidden there in 1837. The low-lying verdant farmland surrounding these cliffs makes them awesome sights.

East of Milton Heights lies a quiet, serene town named Milton, situated on what long ago was known as Sixteen Mile Creek. When the first settlers arrived this area was a virgin forest of tall maple, pine, and oak, broken by sparkling streams. Only the Mississauga Indians had fished and hunted the abundant game. When John Graves Simcoe was made lieutenant-governor of Upper Canada he requested surveyor Augustus Jones to map out the trails and districts which eventually became the County of Halton. This area was then called Martin's Mills after Joseph Martin, the first settler to establish local industry and a great admirer of the poet John Milton. It was at Martin's suggestion that a village meeting gave the town its present name. He built a dam on the creek for his grist mill in 1822, and added a saw mill in 1825.

They still talk in Milton of a great flood in the last century which carried away a section of the millpond and with it the home of an old lady, her pigs, and even the lady herself – all of them floating down the swollen stream together. The Sixteen Mile Creek which served the Martin family so well brought tragedy to them also: in 1846 Martin's son John was drowned there.

In 1857 Milton became the county seat and tenders were issued for the construction of county buildings, with instructions to the architects to design a structure having ventilating and heating apparatus, a court room with a judges' chair, jury rooms with one pine table and two benches, and water closets. The warden was to provide stone broken by prisoners sentenced to hard labour. Before the courtyard was built, executions were held in public and sightseers came from far and wide; once in 1858 they covered half an acre of ground. The Milton Town Hall was built in 1865. The basement originally had openings big enough to drive a wagon through, and for a good many years the local market was held under the building.

Several carpenter and cabinet shops operated in the early days of Milton. One cabinet maker, Englebert Bones, advertised that he produced furniture or coffins to order, and supplied a hearse free of charge to those ordering the latter.

One of the earliest business ventures was a forge opened by the Waldie brothers. Their descendants still operate the business in approximately the same location. Not many teams are shod now, but racehorses in the district keep the shop busy.

One of the largest industries in Milton today is the P.L. Robertson Manufacturing Company. Its head, Peter Robertson, once gashed his arm when his screwdriver slipped, and determined to manufacture a screw from which a driver could not slip. After considerable experimentation he produced the socket-head screw, which is used almost everywhere today. Other Milton

Cedar Springs rushes down a crevice in the escarpment near Milton.

industries have produced chemicals, instruments, lumber, metal goods, tools, steel tanks, bronze and aluminum castings, automobile springs, and dairy products.

Milton is now a shopping centre for farming families who take great pride in their fall fair, which draws people from all over the county. In the fair's early days, farm women proudly displayed their home-made wares in an old frame workshop on Martin Street. Draft horses were shown on Main Street, and roadsters were displayed trotting up and down the First Line. Today the fair covers twenty-one acres.

Close by are Campbellville and Lowville. John Campbell spent his first months in the district, in 1832, in a crude shelter made of tree boughs, before erecting a more substantial cabin and then a saw mill, the first industrial project in the area. In time more people followed. Today Campbellville is a quiet community where only the sound of an occasional Canadian Pacific train disturbs the peace.

A few miles south, Lowville nestles in a curve of the escarpment on Bronte Creek. Perhaps the most beautiful church on the escarpment is located here – the Church of St George, begun in 1898. The edifice was constructed by parisho-ners of stone hand-hewn from the escarpment.

Not far away have been found traces of the earliest people in this district. Because gaps in the escarpment were the most likely east-west travel routes for early man, archaeologists have been combing such places from Beaver Valley to the Niagara Peninsula for evidence of Stone Age camp sites. Arrow heads have been found beneath Mount Nemo in the Bronte Creek gap.

In the seventeenth century the land south of Milton, at the east end of Lake Ontario and around Burlington Bay, was inhabited mostly by Indian tribes, chiefly the Attiwandaronk or Neutrals. In 1669 La Salle arrived at Burlington Bay, pulling his nine canoes ashore on Burlington Beach not far from an Indian village, at a place now known as La Salle Park. Their guide disappeared through the nearest gap in the escarpment to fetch Indians to carry their supplies. Galinée, a priest travelling with the explorer, described their stay:

It was at this place, whilst waiting for the principal persons of the village to come with some men to carry our baggage, that M. de la Salle, having gone hunting, brought back a high fever which pulled him down a great deal in a few days. Some say it was at the sight of three large rattlesnakes he found in his path whilst climbing a rock that the fever seized him. ...At last, after three days waiting, the principal persons and almost every one in the village came to find us. We held council in our camp, where my Dutchman succeeded [in translating for us]. We made two presents in order to obtain two slaves, and a third to get our packs carried to the village. The Indians made us two presents; the first of fourteen or fifteen dressed deer skins, to tell us that they were

St George's Anglican Church in Lowville was completed at the turn of the century.

going to take us to their village, but were only a handful of people, incapable of resisting us, and begged us to do them no harm and not to burn them as the French had burned the Mohawks. We assured them of our good will. They made us another present of about five thousand wampum beads, and lastly, of two slaves for guides.

From 1776 to 1812 the Burlington district was settled mainly by United Empire Loyalists. In 1798, the area where Burlington now stands was purchased by the British from the Mississauga Indians. The Crown gave Joseph Brant, chief of the Iroquois, 3,450 acres because of his loyalty to the British cause and his gallantry during the American Revolution, when he commanded Indian troops. He built a large two-storey white house at what was then called Wellington Square. The Brant Museum, a replica of his home, now stands there. As settlers moved into the area and started to farm, they unearthed many burial mounds, complete with skeletons and the possessions of the deceased.

During the War of 1812 there was much fighting in the district. The nearest was at the Battle of Burlington Heights, where the British defeated the Americans. After the war many settlers from the Niagara frontier came to the district and, although the government had granted the territory to the Indians of the Six Nations, the latter were not allowed to keep it for long.

In 1873, Wellington Square was combined with the village of Port Nelson and named Burlington, a town sheltered from winter winds by the Niagara escarpment to the north and west and cooled in summer by the waters of Lake Ontario. While it was still known as Wellington Square, the harbour was the largest grain and timber port in the area. But as the area became settled a canal was cut across the narrow neck of land that separated Burlington Bay from Lake Ontario – and at once trade began to shift to Dundas and Hamilton on the mainland at the head of the bay. In 1823 the canal was deepened to permit the passage of larger vessels, fostering the growth of Hamilton beyond all other communities in the district.

In 1877 the first train of the Hamilton North West Railway carried material for a swing bridge over the Burlington Canal. The bridge was completed in 1879 and first opened for the schooner *Orient*. By 1880 train service was developed and summer resort hotels and cottages sprang up along the Lake Ontario beach.

With the onslaught of the Automobile Age about 1905, when Ontario became road conscious and boat and rail transportation began to decline in importance, new relationships grew up. Many trunk roads were built. Among them was the Toronto-Hamilton Highway, the first concrete road in the province. It changed Burlington into a residential town for people working in Hamilton. Today the old swing bridge has long since gone, and passers-by on the Queen Elizabeth Way catch only a view of the community from above as

The interior of the Brant Museum in Burlington houses memorabilia of the Iroquois chief. Joseph Brant was responsible for establishing peace between white pioneers and Indians, and for organizing the Indians to help the British drive back American forces during the American Revolution.

The Brant Museum in Burlington is a replica of Captain Joseph Brant's original home and is built on the original site.

they speed over the approaches and high structure of the Burlington Skyway which now spans the canal.

The settlement pattern of Upper Canada was determined by three major geographical features: protection from severe weather, soil suitable for farming, and water power for mills. Waterdown, situated on top of the escarpment west of Burlington just north of Dundas, where Grindstone Creek plunges over the Great Falls into the valley, met all three of these requirements. The first land grant there was made in 1796, but it was the Griffin family, who arrived in 1823, who cleared the land, farmed it, and eventually built three mills, including the first woollen mill in Upper Canada. They also mapped out the original village, for which most of the lots were sold by 1851. The mid-1800s saw great industrialization in the Waterdown area, but many of the old factories have burned down or been torn down and only foundations remain where they stood.

The stories of Ancaster, Dundas, and Hamilton form a trilogy of commercial development in the history of the escarpment. Canals, harbours, wharves, mills, roads, and trains gave impetus to the rapid growth of an area which is still expanding.

The earliest settlement of the three was Ancaster. Here Jean-Baptiste Rousseau (1758-1812), interpreter, Indian trader, coureur-du-bois, mill owner, merchant, soldier, and magistrate, brought his family to establish a trading post in the bush in 1795. John Graves Simcoe wrote that Rousseau 'seems, indeed, to be the only person who possesses any great deal of influence with either of those [Indian] nations' in the area, and he proved to be an invaluable government agent in opening up this territory. As the community grew, he first purchased a mill and subsequently took one municipal position after another, until he died after an attack of pleurisy at the front line at Niagara-on-the-Lake, during the War of 1812.

Toward the end of the eighteenth century, Governor Simcoe's offer of free uncleared land to settlers from the United States caused a great upsurge in the Ancaster area population. In 1785 a privately financed road linked the district to the Niagara region but – like most pioneer roads of the time – it was virtually impassable to wheeled vehicles. Tree trunks and stumps were a perpetual hazard to springs and spokes, so travel was mostly by foot until 1800, when a road was opened from Ancaster to Kingston. By 1827, Ancaster was served by a stage line running from St Catharines to London.

The War of 1812 struck Ancaster, as it did most of the southern escarpment, with lost lives, overnight heroes, and tales to be told during many long, cold winter nights. Here, however, a word was added to the vocabulary of the war – treason – and pro-American sympathizers were hunted out. The Bloody Assize, held in 1814 in Ancaster's Union Hotel, tried nineteen American collaborators, convicting fifteen of them to be hanged as a public example. Seven of them were reprieved, banished, and their property confiscated. The

Old Mountain Mill in Ancaster is the only water-powered mill still in operation on the escarpment.

An imported French millstone in the Old Mountain Mill.

public execution of the remaining eight took place at the military encampment on Burlington Heights.

Seven years later, also in Ancaster, a youth named Peter Jones – son of an English surveyor and an Ojibway woman – was won over to Christianity at a Methodist camp meeting. He spent the rest of his life travelling through Ontario converting his people. Eventually he was ordained a minister and wrote, among several books, an Ojibway speller, an Ojibway hymn book, and translations of the gospels of St Matthew and St John in Indian.

For many years Ancaster's industry was based on mills. To prevent the settlers from being cheated, an early legislature decided to limit the millers' fees. An old story tells of a local miller who was consulted as to what fee would be appropriate. The original sum was to be one-tenth of each settler's grain that was ground, but the miller, whose mathematics were as slow as his millstones, insisted that this was not enough and that one-twelfth seemed more reasonable. His suggestion was made law, but in spite of it Ancaster prospered. Rousseau's first mill has long since disappeared but the Ancaster Mountain Mill – built about 1863 near the original site – continues in use today, and is said to be the only water-powered one still in operation on the escarpment. The flow from its millpond, passing through flume and water wheel, turns the maple wood gears and half-ton millstones.

On January 27, 1891, a famous and brutal murder took place in Ancaster. During the night two masked men battered down the door of Municipal Treasurer John Heslop's home. Heslop was awakened by the noise and, grabbing a small bedside chair, rushed to the head of the stairs. The intruders demanded the municipal tax money. Heslop refused and hurled the chair at them. A shot was fired and Heslop fell to the floor dead. The two men entered his room and ordered his wife to open the safe. When she refused they broke it open but found for their murder only valueless documents. They then ran out of the house and drove away in a carriage with two accomplices. Only tracks in the snow remained by the time help arrived.

The murder remained a mystery until the end of the summer, when three men were charged. A fourth man, already in jail for theft, was also charged. When the Wentworth spring assizes opened in March 1892, and the trial began, the prisoners, Samuel Goosey, John Bartram, John Lotteridge, and George Douglas tried to pin the guilt on one another.

The trial was one of the great social events of the season. Clothes were torn and hats were crushed in the struggle for seats and eventually tickets to the court house had to be issued. The highlight of the trial was a declaration by Goosey that Bartram had fired the fatal shot. The men gave such contradictory evidence, however, that the jury, after three hours of deliberation, brought in a verdict of not guilty. The murder remained unsolved. John Heslop's daughter is reported to have said that the tax money was hidden in the potato bin by her mother. The Heslop house is now known as Woodend and has become the office

Woodend, at Ancaster, scene of the famous Heslop murder, is now being restored by the Hamilton Region Conservation Authority.

At Webster's Falls above Dundas the waters of Spencer Creek tumble over the escarpment.

Coote's Paradise, a swamp named in pioneer days for its abundance of wild ducks, stretches from the edge of Dundas to the escarpment in the background. Pilings of the old Desjardins Canal may be seen in the distance.

of the Hamilton Region Conservation Authority, which is undertaking its restoration.

Just north of Ancaster, two miles west of Hamilton, lies Dundas, now a suburb of Hamilton. In 1787 Anne Mordens settled with her children and grandchildren on the site of Dundas, or Coote's Paradise as it was then called, after the nearby marsh which was notable for its plentiful wild fowl. Other families slowly followed. In 1810, as the pressures grew for a new district seat, Dundas was a close rival for Ancaster. Passions ran high and name-calling became quite common. What use, said Ancastrians, was Dundas, sitting at the end of a long frog marsh which was navigable only in some seasons of the year? But in the end both towns lost out to the yet-unborn Hamilton.

Dundas was far from useless, however. Spencer Creek, a fine mill stream, became its industrial centre during the early 1800s. A three-mile stretch of the stream above Webster's Falls contained no fewer than eight dam sites whose mills produced the usual grist and lumber. Whiskey, barrels, wool, linseed oil, wagon wheels, beer, baskets, axes, and cotton were also made in Dundas. The first paper mill in this province was built there in 1826.

As the local forests were cut down, spring floods and summer droughts increased. Many mills closed down as a result and Dundas struggled to maintain its position as the busiest port at the head of the lake. In 1820 Peter Desjardins persuaded the government to give him money for a canal for water traffic, to replace the winding creek through the marsh then in use. The government grabbed the chance to solve local problems and, twelve years and many more grants later, in August 1832 the canal was opened. It proved a busy waterway for many years, as big paddle-wheel boats carried produce from the Dundas wharves. Then slowly it began to silt up, and with lower water levels traffic gradually declined. In desperation a channel was cut through Burlington Heights in the 1850s in the hope of raising the water level, but to no avail. The canal closed, and from that point onward Dundas lost to Hamilton as a manufacturing centre and port. The days of glory were over; Desjardins' Canal had become Desjardins' Ditch.

Dundas' more famous residents included the rebel leader William Lyon Mackenzie, who operated a drug store there in 1820. Sir William Osler, the famous nineteenth century professor of medicine who revolutionized the treatment of patients in hospitals, lived there as a child and during his early career replaced the local doctor. In its hey-day, Dundas went on a building spree. Architect Francis Hawkins of Dundas designed the Roman Classic style stone town hall, estimated to cost £2000. James Scott, a local builder, completed the hall in 1849 and the building stands today little altered since that time, but now merely a reminder of the past.

Both Dundas and Ancaster have been overtaken as commercial leaders by the industrial giant next door – Hamilton, the divided city.

The original town hall at Ancaster, built in 1871, was the scene of many community gatherings, such as harvest suppers, travelling shows, local fairs, church socials, and elections.

Union Cemetery in Dundas, dating back to 1820, became a park in 1921. The old gravestones were mounted to form this cenotaph.

Part three The South Section

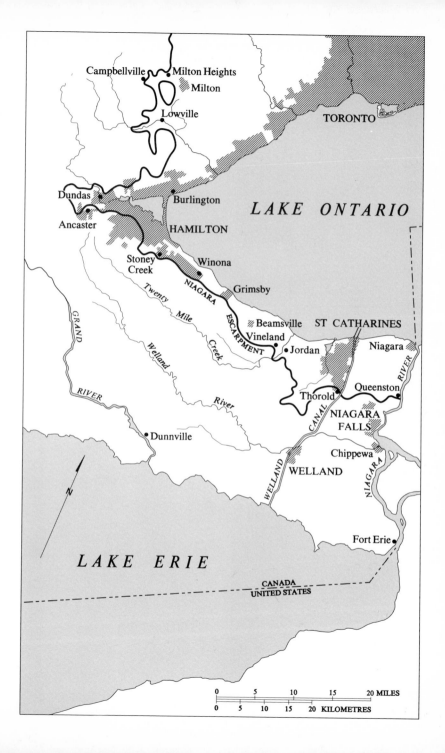

Part three The South Section
Hamilton to Niagara Falls

Stretching east from Grimsby Heights is the densely populated rim of Lake Ontario, which supports a mixture of heavy industry, fruit growing, and truck farming beneath the shadow of the escarpment.

It was almost 6 before we reached the 20 mile Pond, the Mouth of another
Creek. A small Inlet from the lake carries you into this pond which is two miles
long. The banks are very high, of a fine Verdure, & the summits covered with
Wood which was now reflected with the deepest Shades in the Water & had a
most beautiful appearance, which was soon heightened by the rising moon
giving more force to the Shades. Two Houses of Coll. Butler's were distinguished at a distance. . . . while some salmon we bought of an Indian as we
passed Burlington Bay was preparing for our Supper, we walked half a mile
with the Children to a farm House, which we found inhabited by some
Pennsylvanians whom Gov. Simcoe had assisted last Year at Niagara; we had
here excellent Bread & Milk & Butter. We then returned to the tents
& . . . supped by Star light amid this fine scenery of Wood and Water. The bright
fires of the Soldiers below the hill, contrasted with a dark sky now & then
brightened by a gleam of Moon light, had a beautiful effect.
from *The Diary of Mrs John Graves Simcoe*
May 10, 1794

The southern region of the Niagara Escarpment has had the longest history, for
it was the first to be settled and flourishes still under the mixed blessings of the
industrial age. The first settlers were United Empire Loyalists who fled during
and after the Revolutionary War in the United States, deprived of all their
property. In compensation the British government offered them land grants in
this region. Here they settled along the streams which tumbled down the
escarpment and which they named with numbers according to their distance in
miles from the Niagara River. The early political and social history of Upper
Canada began here as the towns and villages of the Niagara Escarpment sprang
up, competing for industrial and political dominance.

The Niagara Escarpment – 'The Mountain' it is called locally – divides
present-day Hamilton in half. Beneath its rim, great steel industries belch forth
clouds of smoke, steam, and fire. Towers of glass and steel rival its height.
Gray, dirty waters of Hamilton Harbour wash against wharves where ships from
around the world unload cargo – at a spot which centuries ago the Indians called
Macassa, 'Beautiful Waters.' The constant cycle of building, wrecking, and
rebuilding gives Hamilton an aura of excitement unequalled elsewhere on the
escarpment. The divided city is a place of contrast: the elegance of Dundurn
Castle, the brashness of the waterfront, the classic beauty of Whitehern, and the
modern symmetry of the new theatre complex.

The tombstones of two men, Richard Beasley and Robert Land, bear the
claim 'the first settler at the head of the lake'; and two other men, Charles
Depew and his brother-in-law George Stewart, pulled their canoe onto Burlington Beach around the same time as Beasley and Land arrived. Local records are

The tall buildings of the business section of Hamilton crowd the cliffs of the escarpment.

Beside what Indians once called 'Beautiful Waters,' Hamilton's blast furnaces singe the sky.

inconclusive as to who should have the honour, but Land is generally credited as the first settler.

Land had come to the New World from England to settle on the Delaware River in 1750. When the Revolutionary War broke out he attacked the colonial cause and became a despatch carrier for the British forces. While he was away on duty one night, an Indian awoke his family with the words, 'House burn. Get children out.' Soon the tearful family were watching their house blaze at rebels' hands while they hid helplessly in a cornfield. They fled to New York immediately and put themselves in the hands of the British forces. For months Land had no idea where his family was; his search led to capture and imprisonment before he managed to escape to New York himself. There the family was briefly reunited, but soon he was a despatch rider for the army again. This time his family was removed with the rest of the Loyalists to Halifax and finally to New Brunswick, but Land was never told. In despair at their second disappearance he moved to the head of Lake Ontario and in 1882 started a new life there as a frontiersman and trapper. Later his wife and children decided to move to Quebec via New York, and en route were told of a man with their family name who had gone to Canada. Full of hope, they followed his trail up the escarpment and caught sight of him at his daily chores, nine years after he had established his lonely homestead.

During the years after Land's arrival, many others settled in the area. They soon tired of trekking either to York or to Niagara to conduct business, and petitioned the government to create a new district at the lakehead. Each established community – Dundas, Ancaster, Stoney Creek – clamoured to be the district seat. As well, a farmer named James Durand petitioned the Assembly to make his land the proposed town's site. His seemed a logical location; the gently sloping plain from the escarpment's base to the lake was the meeting place of trails from Niagara, York, and Brant's Ford. But the War of 1812 postponed the decision, and Durand sold out to the merchant George Hamilton, eldest son of the merchant prince of Queenston, Robert Hamilton. In 1813 Hamilton had the land surveyed into four blocks of town lots. Three years later the new district seat was situated there, and named after him. For years afterward, newly-arriving settlers repeatedly uncovered old Indian camp grounds and burial sites on the farm, complete with artifacts and remains.

Unlike those of other cities, many of Hamilton's early Victorian buildings can still be seen because they were made of durable stone from the mountain. One of the landmarks of Hamilton is Whitehern, a stately twenty-four-room Georgian mansion built in stone by Richard Duggan in 1843. The house was bought in 1852 for the sum of £800 by Calvin McQuesten, who with his cousin, foundry-owner John Fisher, manufactured Ontario's first threshing machine. Today, given to the city by the McQuesten family, it adorns the new civic square as a museum furnished as the family left it, full of treasures.

8

Whitehern, the stately Georgian townhouse of the McQuesten family, has a flat-roofed Ionic porch with a Renaissance balustrade and a pair of French doors leading onto the porch roof.

In the drawing room at Whitehern most of the opulent Victorian furnishings remain very much as they were in Calvin McQuesten's day.

The dining room at Whitehern.

The man who helped bring industrial might to the city is also perhaps the best known figure in Hamilton's history, Sir Allan Napier MacNab, who in 1826 was the city's first resident lawyer. Soldier, landholder, financier, politician, and ultimately prime minister of the Province of Canada, MacNab wielded influence that routed the Great Western Railway along the Hamilton waterfront, opening the southwestern Ontario market to the city's industry. When he was home, 'The Laird' held court in an elegant stuccoed brick mansion, Dundurn Castle, on Hamilton Harbour. Built in 1835 and designed by Charles Wetherell, Dundurn was the finest house west of Montreal – a strange hodgepodge of Italianate and English country manor. Seventy-two rooms housed the lavish entertainment of friends and dignitaries. But the expense was great; Dundurn was heavily mortgaged and at MacNab's death all the furnishings were sold to pay his debts. The castle served as a school for the deaf until 1899 when the city purchased it and eventually restored it as a museum to the elegance of the early 1850s, when MacNab was prime minister.

Ancaster and Dundas had had a considerable head start over the new town as manufacturing centres, and when Hamilton was young Dundas was the acknowledged port of the lakehead. Indeed, because Hamilton had no source of water power, the town had to wait until the advent of steam before it could flourish. But when MacNab had the proposed railway diverted from Ancaster the future was set, and by 1867 the new railway's own locomotive shops and rolling mills were Hamilton's largest industry. The Gurney Iron Foundry's cast iron stoves and scales, the Sawyer Company's agricultural implements, the Wanzer Company's first Canadian sewing machines, and Sanford, McInnes and Company's textiles added to the city's growing reputation as an industrial giant. Bellows, sails, steel springs, hoop skirt frames, biscuits, tinware, soda water, whips, and vinegar were among the wide variety of products made in Hamilton. It was also the centre for a wide agricultural area.

By the time of Confederation Hamilton was a bizarre town, bursting with daily events which ranged from the horrific through the silly to the sublime. One of the greatest of the horrors struck on a lovely March day in 1857 when the afternoon train from Toronto was approaching the city. Just after the engine entered the bridge which spanned the Desjardins Canal at Burlington Heights, the train's whistle blew, signalling it to stop. As the brakes screeched on, the engine truck's forward axle broke loose; its right wheel left the rail and tore through the bridge timbers, sending the train plunging into the abyss. The engine crew and seventy passengers were lost. The glow of industrial progress abruptly paled. Even after a new and better bridge was built, passengers insisted on walking over it. A monument to the victims of the Desjardins Canal Disaster may be seen in the Hamilton Cemetery.

Hamilton installed the first telephone exchange in the British Empire and the eighth in the world in 1878, when the Hamilton District Telegraph Company

9

The classic lines of the colonnade at the front of Dundern Castle contrast sharply with the Italianate portico at the rear. In its original state, the castle, which for many years was supplemented and remodelled almost continuously, was the finest Regency building in Canada.

The suspension staircase in the main hall at Dundern is made of walnut. It was actually added some time after the building was completed in 1835. By the time McNab had completed his expansion and remodelling of the castle, he is estimated to have spent $175,000.

established lines four years after Bell conceived of the device at Brantford. The following year the city was blessed with mass transit in the form of the Hamilton Street Railway: streetcars pulled by horses.

Electricity produced by five persistent and visionary men spurred Hamilton's prosperity to greater heights and established the base from which the mammoth Ontario Hydro sprang. At the turn of the century, 'the Five Johns,' Dickenson, Gibson, Moodie, Patterson, and Sutherland, developed a preposterous idea – the long-distance transmission of electricity. Because they could interest no public official in the scheme, in 1896 they formed the Cataract Power, Light, and Traction Company, which brought electric power to Hamilton from DeCew Falls, on the escarpment near St Catharines. The power was created using water from the old Welland Canal, brought over the falls to a generating plant at the foot.

The opening of the Welland Canal in the 1880s opened a western market for Hamilton's products. The demand for iron and steel increased and companies merged and grew. In 1910 steel companies from Montreal and Ontario united to form the Steel Company of Canada. The large factories of this company – the newest of them highly automated – today stretch along the Hamilton waterfront, crowning the city with an orange glow at night. Since threshing-machine manufacturer John Fisher opened the first foundry in 1835, the industry has grown to produce five and one-half million tons of steel annually in sheets, bars, rods, and wire that are shipped across Canada and exported to more than fifty countries around the world. In addition, Stelco is spending millions of dollars to fight the gray pall of pollution that too often accompanies prosperity.

From almost anywhere in the city, two incline railways could once be seen crawling up the face of the escarpment. The first, built in 1892, was at the head of James Street. The second, built in 1899, was at the head of Wentworth Street in the east end. There was also a flight of three hundred wooden steps leading up the escarpment, but most people paid the nickel fare for the short, easy ride. Later, the only incline railways on the escarpment disappeared with the arrival of modern transportation. All that remains are the scars on the face of the mountain where the tracks once climbed.

Further east is Stoney Creek, where many hundred United Empire Loyalists, fleeing from the American revolutionary forces, settled in the shelter of the mountain, amidst rich farmlands, wooded slopes, and plentiful water. When Ancaster and Dundas lost the district seat to Hamilton, Stoney Creek had been an equally strong contender. The scattered population of the district of Saltfleet was consolidated there, and for a time the town was the region's port and trade centre. Docks and warehouses handled one hundred thousand bushels of grain per year and rivaled Hamilton as the entrepôt of Lake Ontario. Other small communities provided a ready market for the district's farm produce. But when the railroad arrived in Hamilton in 1853, Stoney Creek – like the other

The new Medical Centre at McMaster University reflects in its architecture the giant industrial complexes of today's community, and the need for forward-looking, adaptable construction in the health sciences.

At Canada's largest steel company in Hamilton, a ladle has just been filled with molten steel from the pouring spout of a large open-hearth furnace.

A hectic moment on a blast furnace pouring floor.

rivals – was finished as a commercial base. Most of the warehouses closed or moved to Hamilton. The district returned to farming and fruit-growing, which fostered today's cannery and wine industries.

Today Stoney Creek is best known as a battleground of armies rather than of merchants. The Americans had captured Fort George at Niagara, one of the decisive battles of the War of 1812, forcing the British troops under General John Vincent to withdraw along the escarpment's plain. But Vincent could not retreat, for he was under orders to keep open the lines of communication between Kingston and the army at Detroit. He tried to reach Burlington Heights in order to regroup and make a final stand. The American army of three thousand men outnumbered the British four to one, and the Yankees were determined to cut this vital line and force the British into an untenable position.

The confrontation came on June 5, 1813. The American forces arrived at Stoney Creek and camped at the Gage farmhouse, confident of a victory in the morning.

One of the heroes of the day was nineteen-year-old Billy Green, who had grown up in the area and hunted there, and knew every inch of the territory well. This is his version of what happened, reprinted long afterwards in the Hamilton *Spectator*.

We heard that the Americans were camped down east below the forty so my brother Levi and I went down the road on top of the Mountain. ... We got to the forty and stayed out on the peak of the Mountain ... until noon, when we espied the troops. ... We stayed there until all the enemy but a few had passed through the village. Then we yelled like Indians. I tell you those simple fellows did run. Then we ran along the Mountain and took down to the road that the Americans had just passed over. Levi ran across a soldier with his boots off. ... The soldier grabbed for his gun, but Levi was too quick for him, hitting him with a stick until he yelled and some of the scouts fired at us.

We made our way to the top of the Mountain again. ... [We] went to brother Levi's place on the side of the Mountain. When we heard [the enemy] coming through the village of Stoney Creek, we all went out on the brow of the hill to see them. Some of them espied us and fired at us. One of the balls struck the bars where Tina, my brother Levi's wife, was sitting holding Hannah, her oldest child, on her arm. We all went back to the Mountain to one of Jim Stoney's trapping huts. ...

Not long after, two American officers came up to the house and asked [Tina] if she had seen any Indians around there. She said there was a band of Indians on the Mountain. The officers left, and Tina came out to where we were hiding and whistled. I answered her and told them I would go down to Isaac Corman's. When I got there I whistled and out came Kezieh, my sister. I asked her where Isaac was and she said the enemy had taken him prisoner

At Stoney Creek, a monument to those who fought in 1813 was erected a century later in Battlefield Park.

The James Gage farmhouse became the American force's headquarters during their encampment at Stoney Creek; the family was locked in the cellar and outbuilding. It was located in an ideal military position, with a swamp on the right, the steep bank of the creek in front, and the escarpment on the left. Here Mary Gage nursed the wounded of both sides in 1813.

and taken the trail to the beach. ... I started and ran; every now and then I would whistle until I got across the creek. When I heard Isaac's hoot like an owl, I thought the enemy had him there, but he was coming back alone. ...

Isaac gave the [secret American] countersign to me; I got it and away I came. When I got up the road aways I forgot it and I didn't know what to do; so I pulled my coat over my head and trotted across the road on my feet and hands like a bear. I went up the road to Levi's house and got Levi's old horse 'Tip' and led him along the mountainside until I could get to the top. Then I rode him away around the gulley, where I dismounted and tied old Tip to the fence and left him there, making my way on foot to Burlington Heights. ...

After a hurried interview with the commander, that night the nimble country boy guided the British over the rough terrain to the battleground, complaining all the way that the regulars could not keep up. Once at Stoney Creek, Billy remembered the countersign, tricked the sentries, and led the shouting British into a campground loud with snores. The surprise British attack by night captured both enemy artillery and the commanding general. While the British secured the high ground, the Americans retreated to Grimsby; their camp was finished off by a bombardment from British ships on Lake Ontario under the command of Sir James Yeo. The Stoney Creek region thus was saved for Canada.

Battlefield House, still standing in the middle of the old American encampment, was the home of Mary Gage and her family and was built late in the eighteenth century. The original house was built of logs, but was replaced soon after by a one-and-a-half-storey frame farmhouse. Purchased in 1890 by the Women's Wentworth Historical Society, and subsequently renovated and refurnished for public viewing, it is the only historic site owned and operated by a patriotic society in Canada. The tall monument on the hill behind the house was erected by the government and unveiled on the hundredth anniversary of the battle.

In 1899, Allen Smith, while plowing a field, uncovered a button, and later bits of uniform and a bone. Thus the burial ground of the battle was discovered, after eighty-seven years. In 1908 a cairn of boulders and stones was erected on the site, incorporating a flag with a red field made from the local red clay.

Stoney Creek is also the birthplace of the Women's Institute, an organization devoted to the encouragement of good homemaking. The charter was drawn up in 1897 by Adelaide Hoodless in the home of Erland Lee, situated on the escarpment above Stoney Creek. The institute spread around the world, and today boasts over 37,000 members. Stanley Baldwin, the pre-war prime minister of Great Britain, once said that the Women's Institute was Canada's greatest gift to the British Empire.

Close to Stoney Creek a salt industry developed in 1812 and flourished for three-quarters of a century. Because salt was then one of the principal methods of preserving food and was frequently in short supply, the several excellent salt springs in the vicinity spawned a thriving industry, initiated by Alan McDougall. McDougall was soon joined by William Kent, who with E.C. Griffin built Salt Works Farm. With a 400-foot salt well and fifty salt kettles, the farm monopolized the local salt market and pushed prices up drastically – to ten dollars a bushel and fifty dollars a barrel – until 1883, when imported American salt became readily available. The local industry declined soon afterward, and nothing remains of the old works. But it was from this industry that the township got its name: Saltfleet, or creek of salt.

The village of Winona, which was called Ontario before the province was so named, also grew up during the 1812 period. Legend has it that the settlement was renamed for Tecumseh's daughter, the Princess Winona, who is said to have jumped to her death from the edge of the escarpment above the townsite. A fine Georgian manor in the town was built by Levi Lewis in 1843 from local materials. The bricks were kilned on the property and much of the woodwork in the house came from the locality. The house is still owned by the Lewis family, who run an antique shop from the old kitchen.

Forty-Mile-Creek, or the Forty (renamed Grimsby in 1816) is perhaps one of the oldest settlements of the Niagara region. The first record of the United Empire Loyalists there is dated 1787, but some of the Loyalists must have been there still earlier, and much of the Indian land in the area was swiftly bought by the government for the new arrivals from the rebel states. By 1790 there were enough people to hold a town meeting in John Green's house to deal with such matters as fence heights and livestock registration. Crown-appointed authorities still controlled more important matters, but this was the beginning of local self-government in the province. With the arrival of John Graves Simcoe as lieutenant-governor of Upper Canada in 1792, the District of Nassau (most of the Niagara Peninsula) was divided into counties, and townships were named instead of numbered. Simcoe chose the names of familiar shrines and villages in England, or, as in the case of Dundas, the names of close friends. Township number six became Grimsby Township. Mrs Simcoe thus described the settlement in 1794: 'The mouth of this creek forms a very fine Scene ... Some Cottages are prettily placed on the banks of the River, and a saw mill affords a quantity of boards which piled up in a wood makes a varied foreground.' During the War of 1812, it was here that the Americans, retreating from Stoney Creek, were bombarded by the British fleet on Lake Ontario. The shelling was followed by a militia attack which forced the enemy back to Fort George.

Colonel Robert Nelles, a Loyalist from the Mohawk Valley of New York who served with the Department of Indian Affairs during the revolution, moved to the Grimsby area with his father and younger brothers in the 1780s. The Nelles home, known as the Manor, was begun in 1788 and took ten years to

The Erland Lee homestead on the escarpment at Stoney Creek, where a group of Ontario women signed the founding charter for the Women's Institute. Since then rural women throughout Canada and in other countries have organized in the w.i. *and its sister institutes to further home management, agriculture, good citizenship, and cultural activities.*

The pre-Confederation Georgian style brick home of Levi Lewis at Winona is notable for its classical two-storey porch.

The antique shop in the old Lewis home.

build. Labour was scarce; the only carpenters were those from ice-bound lake vessels, who during the winter lived on the premises and built the house entirely from local materials. The three-foot-thick walls were made of plastered stone, and lime for the cement was burned on the property. The house has changed direction over the years; the front door used to face the lake, but with the coming of the road from Hamilton to Niagara the opposite side offered a more pleasant entrance. The War of 1812 raged around the house, and escaping British soldiers hid in a secret hide-away in the wainscotting during a brief American occupation. (At the time Nelles was away commanding the 4th Lincoln militia.) At another time, when Colonel Nelles and Joseph Brant were acting as liaison between the Indians and the British forces, the grounds became a huge Indian encampment. In bad weather, the Indians would slip through the door to sleep by the fireplaces, and the family frequently had to step over them to move about; fortunately the Indians always vacated the premises by dawn. The celebration of war's end was held in the third-floor ballroom, for all the buildings at Niagara had been destroyed and the Manor was the only remaining building large enough for the gathering.

The Nelles family documents contain a fascinating record of the daily pioneer life, including such items as a doctor's bill for 'the bleeding of a wench.' Another document indicates that in the early days, when more established households frequently took in new pioneers, a family sheltered at the Manor brought typhoid, which killed one of the Nelles children.

During the Depression of the 1930s the Manor was almost torn down to provide lots for low-cost housing. Only the efforts of Nelles Rutherford saved it from destruction. The house has since been purchased from the family for restoration.

St Andrew's Church, built of logs in 1794 on land donated by Colonel Nelles, was replaced by a frame building ten years later. The present stone structure, built during the years 1823-5, is one of the most impressive on the escarpment. The churchyard contains the graves of many of the Loyalist families who founded the Forty. When a large tree was cut down recently, it was found to have grown entirely around the memorial tablet of one of these pioneers.

The Stone Shop Museum is housed in a stone building that was erected by Allen Nixon on Crown-granted land. At first it was used as a farm workshop, but during the War of 1812 both British and American forces used it as a smithy. The fan-light door was added much later during reconstruction of the building, which in 1963 became a museum housing the Grimsby Historical Society's collection of early implements, glass, and papers.

The first white men to settle here found wild fruit trees in abundance over the whole area from the escarpment to the lake. John and Joseph Brown, who established their farms in the 1850s, are credited with the first commercially grown peaches on the Niagara Peninsula; in 1880, Brown-grown peaches were

The Grimsby Stone Shop Museum houses relics and documents of local history.

From the Bruce Trail on Grimsby Heights, the town of Grimsby can be seen stretching along the escarpment's base.

The grounds of the home of Colonel Robert Nelles in Grimsby served as a battleground and an Indian encampment during the war of 1812, and with the return of peace as the scene of the local victory ball.

served to Edward, Prince of Wales, at a banquet in the Queen's Hotel in
Toronto. The first peach orchards in Grimsby were planted by A.M. Smith and
Charles Woolverton in 1856. Within the next few years orchards extended
along the escarpment from Winona to Niagara-on-the-Lake. The first canning
factory in Canada was built at Grimsby in 1865 by W. Kitchin. Peaches,
cherries, and pears came to constitute the bulk of local industry. Today Grimsby
is changing rapidly, as families sell their land to real estate developers. The old
quiet village, with its fruit wagons and radial cars on the main street, is no more.
Within another generation the fruit orchards may disappear as well.

Just east of Grimsby, the small community of Grimsby Beach was for many
years a popular and fashionable summer resort. People from all over the country
summered here along the shore of the lake. From 1859 until the early 1900s, the
beach was used as well as a camp and meeting ground for the Methodist
Church. Week-long services, picnics, and general meetings were part of August life then. A famous temple was built which was covered with an immense
wooden dome, 122 feet in diameter and 100 feet high. According to early
accounts, it was possible to see to Toronto from its top. All that is left now of the
structure, which could hold 8,000 persons, is a stone marker and the bell which
called people to services. The Stone Shop Museum in Grimsby displays a scale
model of the temple.

The town of Beamsville began in the final years of the 1700s when Loyalist
Jacob Beam was granted acreage in Clinton Township. He cleared the land and
is reputed to have made the first cheese in the region. The Beam house still
stands in the town.

Beam was soon joined by the Mennonite Kulp families and the blacksmith
Isaac Marlatt. A carriage shop, a tin shop, a pottery, and a drainage tile factory
helped to swell the population, but Beamsville missed its chance at major
industry. About 1870 the Harris Foundry, manufacturer of a popular make of
agricultural equipment and the owner of several patents – and the town's principal source of income – was refused a waterline for its plant and a tax adjustment;
as a result Harris moved his factory to Brantford, where he joined Hart Massey
to form the Massey-Harris Company, now the mammoth Massey-Ferguson
Company. Beamsville survived as a market centre for the surrounding countryside.

In the mid-70s Robert Gibson, a stonecutter from Scotland, established the
Gibson Quarry on top of the mountain, and brought new prosperity and skilled
workers to the town. For nearly twenty years, huge building stones rolled down
the mountain on the quarry tramway to the rail station to be shipped throughout
Ontario. Gibson was killed in an accident shortly after the operation got under
way, but he was succeeded by his nephew William Gibson, later a member of
parliament and a senator. His home in Beamsville became part of the Great
Lakes Christian College after his death. The quarry was shut in 1903. The town

Woodburn Cottage in Beamsville was built in 1834 by the Osborne family.

depended thereafter on fruit and such ancillary industries as the Beamsville Basket Company.

In 1896 the electric Hamilton, Grimsby, and Beamsville Railroad provided convenient transportation to Hamilton for agricultural products. It moved both people and produce for thirty years, until the automobile made it unprofitable.

The first Beamsville Country Fair was held in October 1857. That year cattle, grains, dairy produce, handiwork and vegetables were displayed, including a two-foot-long carrot. The fair was broken up by a free-for-all by local youths who filled the air with flying fruits and vegetables until the exhibitors packed up and went home. The annual fair still generates excitement and fun, without going to such extremes.

Vineland, originally settled by Loyalists and Pennsylvania and New York Germans in the 1790s, now supports a large Mennonite settlement which has nurtured some of the finest farms on the escarpment. The Mennonite burying ground, which dates from 1798, is completely surrounded by a stone wall built by Newton Perry for the sum of ninety silver dollars. The walls were topped with wood covered with sheet iron. Following Mennonite tradition, only name, age, and date of death appear on the headstones.

The widely acclaimed Ontario Horticultural Station at Vineland is located just north of the village on the lake. It was for many years the only station of its kind in Canada and is the site of research in practical horticulture and in new varieties of fruits and vegetables. The station has made valuable contributions to soil and orchard management. Its two hundred acres of grounds, with flower beds and rare trees, are open to the public.

The community of Jordan dates back to 1755, when it was called Twenty-Mile Creek. The Loyalists and Mennonites who settled there in the 1790s and early 1800s renamed the creek Jordan River. The Twenty supported a variety of industries along its banks. Most famous were the mills at impressive Ball's Falls, situated on the top of the escarpment above the town. Here the water falls in two steps, ninety and twenty-five feet high, over the edge. Jacob Ball arrived at the spot with forty men and his two sons, George and John; by the time Jacob died early in the nineteenth century, several mills had risen around the two falls. In later years, the complex included grist, saw, and woollen mills, a copper shop, and a general store. It was so important to the British that they stationed a detachment under the command of General Brock's nephew to guard it during the War of 1812. The mills on the ninety-foot drop of the lower falls still stand, but the woollen mill has been torn down. The entire area is today a 200-acre conservation area which includes a museum in the mill, two furnished pioneer log cabins, and a lime kiln, as well as picnic facilities and nature trails.

The great vineyards of the Jordan Wines Company stretch along the base of the escarpment down to the lake throughout the region. Approximately 80,000 tons of grapes are harvested each year. Most of them go to the making of nine

The sparkling plunge of Ball's Falls, in the popular conservation area of the same name, is a goal for both winter sportsmen and summer hikers. Where hot-weather visitors now cool themselves, over one hundred years ago several mills and a commercial complex stood.

The flour and grist mill at Ball's Falls is now a museum of pioneer relics.

million gallons of wine. The 23,000 acres of grapes were harvested by hand until a few years ago; now mechanical pickers straddle the vines and shake the grapes into hoppers.

The Niagara Grape and Wine Festival began in 1952 in St Catharines. It has grown from a one-day celebration into one of the peninsula's finest fall attractions, lasting a week or more, with parades, vineyard tours, wine-sampling, floats, and bands. Along with the Royal Henley Rowing Regatta held in the same city every July, the festival is one of the great tourist attractions of the region.

The first record of settlement at St Catharines is a list of Loyalists who had settled in the district west of Mill Creek by 1787. The earliest of these in the St Catharines area were John Hainer, a miller, and Jacob Dittrick, a soldier, both from New York State. Pursued by Indians and revolutionary soldiers, they went first to Queenston and thence to Twelve-Mile Creek.

St Catharines first achieved note for its mineral springs. Doctor William Chase, proprietor of a local drug store, had watered his hogs at these springs, and had watched them grow fatter and healthier than any others. When the animals were slaughtered he found that their livers were in exceptionally good condition. Experiments suggested that the basis of porcine health lay in the mineral waters: would they not help human health as well?

On the basis of this information Stephenson House and Springbank House, modeled after the great German spas, were built at the instigation of Dr Theophilus Mack for the sole purpose of providing public access to the mineral waters. The springs were soon famous and people flocked to St Catharines. Dr Mack, founder of St Catharines' first hospital, the General and Marine Hospital on Cherry Street, in 1865, as well as the first school for nurses in North America, the still-functioning Mack's Training School for Nurses, used both houses chiefly for his own patients. His fame spread and soon he was one of Canada's leading physicians. When the Honourable George Brown, editor of the Toronto *Globe*, was shot by a disgruntled former employee, Dr Mack was called to Toronto in the hope he could help the dying man.

Another local industry, shipbuilding, was founded by Lewis Schickluna, who had emigrated from Malta in the early 1800s. He first worked in Quebec, where he helped build the *King William*, the first Canadian steamboat to cross the Atlantic, and then in Oakville, where he worked on the *Transit*, later used as a troop carrier during the Rebellion of 1837. After he came to St Catharines in 1836 he established his own shipyard on the old canal and built about one hundred vessels of various kinds, some of them steamships. By 1871 he was head of a four-million-dollar business. The shipyards have long since closed.

In 1850, the first waterworks in the city piped water from Merritton to the town reservoir through pipes made of bored pine logs. When in 1856 fire destroyed the business section of the city, the waterworks helped to save the rest

Grape vineyards cover the countryside surrounding Jordan.

Mechanical grape-pickers are now employed in the vineyards, feeding the wineries of the Niagara Peninsula.

The General Motors engine plant complex is situated below the escarpment at St Catharines.

of the community. The city's first newspaper, forerunner of the present *St Catharines Standard*, appeared in 1826 as *The Farmer's Journal and Welland Canal Intelligencer*, with two avowed purposes: to provide correct and accurate statements about the Welland Canal, and to bring to the attention of the public new plans for the improvement of navigation. In 1862 St Catharines wrested the county seat from Niagara, and had to pay Niagara $8000 in compensation; many of the public buildings date from that time. The modern city of St Catharines – incorporated in 1879 – is one of the largest manufacturing centres on the peninsula. General Motors of Canada operates two large complexes there employing some seven thousand people.

A short distance west of the city the flumes of the DeCew Falls power station run almost vertically down the escarpment. This station was opened in 1898 and supplied power to Hamilton and other communities: to break the general dependence on coal, the power company leased electric motors to manufacturers who would use its service. Brock University, one of the most modern in Canada, rests on the edge of the escarpment overlooking St Catharines. When the university was founded in the early 1960s, it held classes all over the city until its permanent quarters were completed. Its stark concrete towers soar over the escarpment and the city below with the solidity and permanence of the mountain itself.

Merritton, situated between St Catharines and Thorold, is named after one of the early pioneer families in the area, but originally gloried in the name of Slabtown, probably because of the nearby quarries. Merritton and St Catharines were joined in 1877 by an electric railroad, one of the first in the province. The power was obtained from a direct current generator at Lock 12 on the old Welland Canal; when the canal was shut down for cleaning, the cars were towed by horses. Industry today includes pulp and paper mills, quarries, and automotive parts manufacturing.

Thorold is located on the brow of the mountain about three hundred feet above Lake Ontario. Founded in 1788, the town was named for Sir John Thorold, a British MP noted for his opposition to the war in the colonies. In 1853 the Thorold and Port Dalhousie Railway brought prosperity, in the form of grain carried by rail for transshipment to other centres. Thorold is noted today for its pulp and paper industry; towering piles of logs provide newsprint for papers across the continent. But the town's early growth came with the birth of the Welland Canal in 1829. Rough shelters mushroomed then to house workers' families within the town limits, on sites now lined with industry.

The Welland Canal, one of the great engineering masterpieces of Canada, is approximately twenty-five miles long and in its locks lifts the huge lake freighters 325 feet up the escarpment on their way to the Upper Lakes. Historically there have been three canals. The first had locks built of timber and was only eight feet deep; barges and very small vessels were towed through. In

1842, to allow the passage of larger vessels, the second canal was built of stone quarried from the escarpment; locks were made longer, wider, and deeper and reduced in number from forty to twenty-seven. In 1870 a new canal was built with a width of forty-five feet and locks 270 feet long. In 1973, a seven-mile-long by-pass around Welland was completed so that ships would no longer have to pass through the very centre of the city.

Leaving the roar of industry behind, south of the canal the escarpment recovers its rural serenity. Orchards and farms once more dominate the landscape. As the height of the land increases the road drops into the basin below Brock's Monument to the quiet town of Queenston. This community began in the last years of the eighteenth century, although the circumstances are somewhat vague. Robert Hamilton, an empire-building merchant whose connections spread throughout the lower end of the peninsula, is generally recorded as the first settler, although there is evidence that an earlier mercantile concern predated him by several years. With the arrival of Lieutenant-Governor John Graves Simcoe in 1792 the town, originally known as West Landing became the location of the barracks of the governor's new regiment, the Queen's Rangers. The settlement's name subsequently progressed through various forms of Queen's Town to Queenston.

From his quarters downstream at Niagara, Simcoe tirelessly tramped the Niagara region and further north and west on foot, exploring and assessing the territory to carve civilization out of the sparsely settled wilderness. His wife recorded that on his return from a fifty-mile walk to Burlington Bay from Niagara, he reported that 'The shores of the Lake are for a great distance as high as the Falls of Niagara, & several small rivers falling from that height make very picturesque scenes. He was delighted with the beauty of the Country and the industry of the Inhabitants. He lodged every night in Houses where he was accomodated with a clean Room & a good fire.'

Simcoe encouraged and assisted settlers, ordered new roads and public facilities, and spurred the economic development of the region, all the while maintaining a crude facsimile of English upper-class life at Niagara, with balls and banquets and games of cards. His wife set off into the bush on jaunts of her own, recording for posterity boundless descriptions of early Canada in her diary.

The location of Queenston was ideal for rapid development. It was situated below the great falls at Niagara, where the way by water to York and Kingston lay open. Hamilton and other entrepreneurs successfully promoted the construction of the 'west portage' from Queenston to Chippewa Creek above the falls, and as a result goods, produce, and traffic bound to and from the Lake Erie region streamed into the port. Ox teams carried goods from lake to lake. The most unusual passage over the portage was of the *Washington*, a sailing ship built by Americans to ply Lake Erie but subsequently bought by Canadians who

Ships climb and descend the mountain side by side through the twin locks of the Welland Canal.

saw more profits for her in Lake Ontario. She was dragged on runners by horses and oxen to Queenston. The venture was so succesful that during the Rebellion of 1837 a small fleet was brought across to protect the frontier. Horses belly-deep in mud dragged the boats on wagons. This event marked the virtual end of Queenston's shipping, however; the Welland Canal made access to Lake Erie and points west much simpler.

Queenston boasted Upper Canada's first distributing post office, which was established in 1802 under the direction of postmaster and sheriff Alexander Hamilton, son of Robert and brother of the founder of Hamilton. Mail arrived by steamboat for delivery by stagecoach. After his father's house was destroyed during the War of 1812, Hamilton built his own home, Willowbank, between 1833 and 1835. The stately mansion, designed in the then-popular Classic revival style, sports eight fluted wooden pillars which extend the full two-storey height of the house. The pillars were handcarved in one piece from logs from a neighbouring farm.

The beginning in Ontario of the country fair as we know it today was at Queenston in 1799. The emphasis at first was on livestock-showing, for prizes awarded by the community's farmers. Queenston also saw the first railway in Ontario, the Erie and Ontario, built in 1839 to Chippewa around Niagara Falls. Its carriages were horse-drawn because the steam engines of the day could not negotiate the great rise of land. Travel on it was only a summer proposition until 1854, when more efficient steam power was adopted. The tracks were extended to Niagara and Fort Erie. The line has since been absorbed by what is now the New York Central system and is used only for freight. One end of it can be seen at the steamer dock at Niagara-on-the-Lake.

The Battle of Queenston Heights, a decisive victory for Canadian forces during the War of 1812, cost the life of General Sir Isaac Brock, the commander of the British and Canadian military. Twelve years after the battle a 135-foot monument was erected on the edge of the escarpment in memory of the general and his aide-de-camp, John MacDonell (a distinguished lawyer and at the time attorney-general of Upper Canada), and their bodies were moved to a stone vault in the base. But when the column was already more than forty feet high, it was discovered that followers of the rebel William Lyon Mackenzie had introduced a copy of Mackenzie's paper, the *Colonial Advocate*, into its cornerstone. To remove the newspaper, the enraged lieutenant-governor had the column torn down and rebuilt at government expense. It stood until it was severely damaged in 1840 by a charge of gunpowder laid by another rebel, Benjamin Lett. The bodies were this time removed to Queenston, where they remained until in 1853 the present column was built, 196 feet tall and capped by a statue of General Brock. By this time the general and his aide had been moved four times.

The legendary heroine of 1812 was Laura Ingersoll Secord, who, while

Willowbank, Alexander Hamilton's Home at Queenston, built between 1833 and 1835, is considered one of the finest examples of colonial architecture on the continent. Its wooden Ionic columns, topped with splendid hand-carved capitals, rise in front of the Grecian style doorway. To install the columns without damaging the fluting, the builders bored holes at each end and inserted iron bars; the bars were used to lift the columns into place and were then removed and the holes plugged.

The Brock Monument at Queenston towers over the battleground.

The Laura Secord homestead in Queenston, where Laura overheard American officers discussing a surprise attack.

American troops were billeted at her home in Queenston, learned of their plans to attack Beaver Dam. Her historic twenty-mile walk through American lines to warn Lieutenant James FitzGibbon, the officer-in-charge, is familiar to every Canadian school child. (Less well known is the fact that FitzGibbon already had word of the U.S. plans from his scouts.) The Secord house has recently been restored and is open to the public.

A post-war wave of immigration brought to Canada an obscure Scots draper's assistant, William Lyon Mackenzie, to whom much mention had already been made. After setting up shop in Toronto, he moved to Dundas for a few years, then moved again to Queenston where in 1824 he established his own newspaper, the *Colonial Advocate*. In its columns, in incisive, colourful prose he fumed at the political corruption which he felt was keeping his newly chosen land from achieving its full potential, and levelled charges at a privileged clique who dominated the government of the Canadas, particularly Sir Allan Napier MacNab. He moved to Toronto, where his inflammatory prose resulted in an attack by young members of the establishment on the offices of the newspaper. The incident made him a radical hero and he was elected to the Assembly: here his rhetoric got him expelled several times, but each time won him re-election . He was also elected Toronto's first mayor. His rebellion against the government in 1837 was lost almost before it got started, and Mackenzie fled to the United States, taking refuge in homes and natural hideaways all along the escarpment as he eluded government forces. Although his attempt failed, it provided the impetus for far-reaching government reform.

The hydro-electric development at Queenston is one of the largest in the world. The Sir Adam Beck Niagara Generating Station Number One was opened in 1917 and named after 'the Father of Hydro.' Beck had in 1903 been appointed to investigate the development of electric power from Niagara Falls; in 1906 he introduced to the legislature the bill which created the Hydro-Electric Power Commission of Ontario, of which he was chairman until his death. In 1951 a second intake was begun at Chippewa, increasing generating capacity to about two-and-one-half-million horsepower. Its twin canals carry the water in tunnels under the city of Niagara Falls to the generators at Queenston.

The first Queenston-Lewiston bridge across the Niagara River and the international border opened for traffic in 1851, replacing a ferry service which had been initiated by Governor Simcoe in the late 1700s. The 1,045-foot span supported a twenty-foot roadway atop stone towers. In 1864, the bridge was swept away after the cables had been loosened to permit the flow of ice and were never retightened. The towers were all that remained until 1898, when a new bridge was built.

The escarpment's winding path leaves Ontario at the thundering white water of Niagara Falls. The falls are a relatively recent geological formation, not more

At the bank of the Niagara River the escarpment passes out of Ontario into New York State.

than twenty thousand years old. They were formed when the retreating glaciers of the last ice age altered the topography of the land and forced Lake Erie to overflow into Lake Ontario over the Niagara ledge. In more recent history the area surrounding the falls was for many years the domain of the Neutral Indians, and there are still signs of their encampments along the river. Rumours of the thundering waters reached Jacques Cartier as early as 1535, but the first written record of a visit to the falls was made by Father Louis Hennepin during his travels with La Salle in 1678. The English edition of his book, *A New Discovery of a Vast Country in America*, appeared in 1698 with the following description:

Betwixt the Lake Ontario and Erie, there is a vast and prodigious Cadence of Water which falls down after a surprising and Atonishing manner, insomuch that the Universe does not afford its Parallel. ...The Waters which fall from this horrible Precipice, do foam and boyl after the most hideous manner imaginable, making an outrageous Noise, more terrific than that of Thunder.

La Salle later built a fort on the site, which was eventually captured by the British in 1759 under the command of Sir William Johnson. Johnson's wife, born Molly Brant, and her brother, Captain Joseph Brant, the Indian chief, lived at the falls during the American Revolution, when their influence with Six Nations Indians was of great help to the British cause.

For years there was no permanent settlement at the site of the falls. Many people came to visit and look at this natural wonder, including Mrs Simcoe, who had ladders built at various perilous perspectives, and Nathaniel Hawthorne; but no one stayed. In time a group of mills, the Bridgewater Mills, were erected at the head of the upper rapids, but they were burned by the Americans in 1812 and only some were rebuilt. These were later expanded, and soon a small settlement sprang up, renamed Falls Mills. Because it was so close to the falls, however, all goods had to be carted to it overland, a hard and expensive method of transportation in those days of poor and rough roads. The last of the mills burned in 1870 and was not rebuilt. The property was later sold to the Queen Victoria Niagara Falls Parks Commission, now known simply as the Niagara Parks Commission.

In 1848, the first bridge was constructed across the river gorge. Samuel Zimmerman, owner of great tracts of land in the area, began to plan streets for a new town, and lent his influence to plans for a larger double-decked bridge, built in 1853 to carry both trains and private vehicles between Canada and the United States. Designed by a German engineer, John Roebling, this was the first successful railway suspension bridge in the world. The first locomotive crossed in 1855.

In 1853 the settlement planned by Zimmerman was incorporated as a village named after the then governor-general, Lord Elgin. The rapidly expanding

The whirlpool rapids at the bottom of the escarpment were a favourite haunt of
Mrs John Graves Simcoe. She called it 'a very grand scene . . . where the
current is so strong that Eddies are formed in which huge timber trees carried
down the falls from a sawmill upright.' Later, the rapids became the stage of
daredevils, and it is said that illegal immigrants to the U.S. were smuggled over
the rapids to the opposite bank for a hefty sum – only to find that they were still
in Canada.

Sir Harry Oakes' Tudor edifice on the Niagara River is similar to the homes he owned in Kirkland Lake, Bar Harbour, London, Sussex, Palm Beach, and the Bahamas. After his death Lady Oakes donated the building to the federal government, which eventually returned it for lack of a permanent use for it. It served as a hospital for a few years during and after the Second World War.

community amalgamated with neighboring Clifton three years later. But Zimmerman's death in the Desjardin Canal Disaster in 1857 slowed community growth. It was not until 1881 that that town was renamed Niagara Falls.

By that time it was a brassy and raucous tourist centre. Taverns, museums, and shops lined the river bank, and con-artists of every hue tried to bilk the sightseers. Beginning in mid-century, stunters of all kinds vied to perform daring deeds at the falls. The first such attraction had no heroes and many victims, however: in 1827 an old schooner was sent over the falls loaded down with animals. Two years later the notorious Sam Patch leaped over the falls – twice. But it was not until 1859 that the master of them all, Jean Francois Gravelet, better known as Blondin, for the first time walked a tightrope across the gorge. His repeated efforts to top his previous achievement saw him make the journey at various times on stilts, performing acrobatics, cooking his dinner midway, and carrying a man across on his back. His immense success inspired many pranksters and competitors to seek fame and fortune at the falls, but none surpassed him, not even the woman tightrope-walker Maria Spelterini. In 1883 Captain Matthew Webb died swimming the rapids. Three years later the era of the 'barrel-cranks' began with Carlisle D. Graham's tumble over the falls in a barrel; three were killed, but 'Red' Hill and others gained fame of a kind. In 1911 the first stunt pilot at the falls, Lincoln Beachey, flew his plane under the suspension bridge.

The second bridge over the gorge had proved inadequate to the stress of traffic over it by 1887 and had been remodelled – and enlarged once more before the end of the century, this time incorporating a steel arch. The success of the lower bridges created a demand for a bridge closer to the falls. In 1867 construction was begun on a Falls View Bridge. It was only ten feet wide, restricting traffic to one carriage at a time, and it swayed and rocked in high wind until in 1889 a gale tore it from the bank and sent it plummeting into the river below. Its more famous successor, the Honeymoon Bridge, met much the same fate in 1938: an unusually heavy ice build-up severely damaged its supports and left the bridge hanging on the verge of collapse for several days before it too fell. The present Rainbow Bridge was completed in 1941.

The home of Sir Harry Oakes, Oak Hall, is the most distinguished building near Niagara Falls. The property originally belonged to Colonel Thomas Clark of the 2nd Lincoln Militia, who built a forty-room mansion on it after the War of 1812. This house was largely rebuilt by the land's second owners, the Street family. Oakes, a gold prospector who struck it rich in northern Ontario and became a millionaire, acquired the property in 1924, along with about six thousand acres along the river, on which he began constructing two public golf courses and a bird sanctuary. Here he built a Tudor edifice similar to those on others of his properties in England, Maine, Florida, and Kirkland Lake, Ontario. In 1935 he left Canada, reputedly over a dispute about income tax, and

settled in the Bahamas. His yet unsolved murder there made world headlines in
1943.

One of the most popular trips at Niagara Falls today is the voyage on the *Maid of the Mist* to the base of the Horseshoe Falls. The first steam-powered *Maid of the Mist* was launched in 1846. It was a rather cumbersome vessel of one hundred tons, which was replaced a few years later. After several years that ship was retired in 1860 and made its last journey down the rapids, the first vessel to reach Queenston by water from the west. Two further replacements, built in 1885 and 1892, plied the treacherous waters until 1955, when they were destroyed by fire and replaced by the present vessels.

In the early years of tourism, a huge promontory called Table Rock afforded the finest view of the Falls, until it gradually tore away from the face of the gorge and crumbled into the river. Many artificial viewing structures have been erected for the benefit of visitors, beginning with the Terrapin Tower of 1853 and culminating in the slender observation towers along the Canadian bank today. For a long time it seemed that everything was being built there except decent accommodation; the town had a giant carnival atmosphere which forced the more genteel class of tourist to mingle with a new and commoner breed, much to the chagrin of the former. More and more people also were upset by the fact that this natural wonder was the victim of private exploitation along every inch of the gorge's edge, but nothing was done about it for some time.

In 1878 Lord Dufferin, then the governor-general of Canada, first suggested public parks in the region. Nothing happened until 1885, but then the Ontario government purchased Queen Victoria Park, and thus began the now elaborate holdings of the Niagara Parks Commission. Today miles of clipped lawns and wooded knolls along the river line the approach to the majestic curve of the Canadian Falls. Niagara is a far cry from what it was in Mrs Simcoe's day, but her enthusiasm for the falls – and for the escarpment which forms them – must strike a sympathetic chord in every visitor:

> These scenes have afforded me so much delight that I class this day with those in which I remember to have felt the greatest pleasure from fine objects, whether of Art or Nature.

The waters of the Niagara River slide toward their gigantic plunge over the escarpment at Niagara Falls.

Afterword

The Niagara escarpment should belong to everyone. It is already too late to preserve the natural beauty of many areas on the escarpment which have been ravaged by industrialization and urbanization. But there are great long reaches between Burlington Heights and Tobermory where the escarpment is still in its natural state, and where the rivers and the land have not been polluted. It is still possible to ensure that future generations can enjoy the woods, hills, ravines rivers, waterfalls, and wildlife of one of North America's unique geographical features.

THE AUTHORS

Index